Fostering Transformative Change for Sustainability in the Context of Socio-Ecological Production Landscapes and Seascapes (SEPLS)

Maiko Nishi • Suneetha M. Subramanian •
Himangana Gupta • Madoka Yoshino •
Yasuo Takahashi • Koji Miwa • Tomoko Takeda
Editors

Fostering Transformative Change for Sustainability in the Context of Socio-Ecological Production Landscapes and Seascapes (SEPLS)

 Springer

Editors
Maiko Nishi
United Nations University Institute for the
Advanced Study of Sustainability
(UNU-IAS)
Tokyo, Japan

Suneetha M. Subramanian
United Nations University Institute for the
Advanced Study of Sustainability (UNU-IAS)
Tokyo, Japan

Institute for Global Environmental Strategies
(IGES)
Hayama, Japan

Himangana Gupta
United Nations University Institute for the
Advanced Study of Sustainability
(UNU-IAS)
Tokyo, Japan

Madoka Yoshino
United Nations University Institute for the
Advanced Study of Sustainability (UNU-IAS)
Tokyo, Japan

Institute for Global Environmental Strategies
(IGES)
Hayama, Japan

Yasuo Takahashi
Institute for Global Environmental
Strategies (IGES)
Hayama, Japan

Koji Miwa
Institute for Global Environmental Strategies
(IGES)
Hayama, Japan

Tomoko Takeda
Institute for Global Environmental
Strategies (IGES)
Hayama, Japan

ISBN 978-981-33-6763-0   ISBN 978-981-33-6761-6   (eBook)
https://doi.org/10.1007/978-981-33-6761-6

This Springer imprint is published by the registered company Springer Nature Singapore Pte Ltd.
The registered company address is: 152 Beach Road, #21-01/04 Gateway East, Singapore 189721, Singapore

Foreword

As part of our mission to advance global efforts towards sustainability, UNU-IAS has been contributing to the sustainable use of biodiversity at all levels, through research, policy engagement, and capacity building initiatives. We have focused in particular on promoting resource-management approaches and positive human–nature interactions in various geographically distinct *socio-ecological production landscapes and seascapes* (SEPLS). Integrated approaches for SEPLS management have the potential to solve various local and regional problems while contributing towards achieving global goals for sustainable development and biodiversity. In fact, many case studies show how SEPLS management has helped to improve the quality of life and the environment with multiple benefits, making SEPLS one of the best options for facilitating transformative change. Recently, actions towards such change have been considered promising strategies to deal with myriad challenges of the environment and sustainable development and thus have become an important part of today's global agenda. In particular, transformative change is urgently needed as the focus shifts towards building back better from the COVID-19 pandemic. UNU-IAS is strengthening our role in this effort, working on elements of transformative change in the context of SEPLS.

UNU-IAS has worked closely with the Ministry of the Environment of Japan to develop the Satoyama Initiative, a global effort to realise "societies in harmony with nature". We have hosted the Secretariat of the International Partnership for the Satoyama Initiative (IPSI) since its establishment at the Tenth Meeting of the Conference of the Parties to the Convention on Biological Diversity (CBD COP 10) in Aichi, Nagoya, Japan, in 2010. In this role, we coordinate the efforts of partners across the globe towards biodiversity conservation through integrated and holistic landscape and seascape management approaches. IPSI is celebrating its 10th anniversary in 2020, having accumulated a diverse range of knowledge and experience, including a database of 198 case studies submitted by member organisations. The membership of IPSI has grown from 51 to 271 organisations over the decade, ranging from governmental, non-governmental, and research organisations to indigenous peoples' organisations, working collaboratively for better management of SEPLS in various settings around the world.

The Satoyama Initiative Thematic Review (SITR) has been published since 2015 to share this knowledge and showcase the diverse work of IPSI. This sixth volume presents case studies that demonstrate endogenous initiatives to cope with emerging environmental challenges and adapt to changes at the local and regional levels, ultimately leading to *transformative change* for sustainability. Transformative change is an emerging concept, which has been featured in the latest global assessment of the Intergovernmental Science-Policy Platform on Biodiversity and Ecosystem Services (IPBES) among others, and is central to the Post-2020 Global Biodiversity Framework of the CBD. However, further studies are needed on its processes and mechanisms. Relevant knowledge will be assessed through the upcoming IPBES Transformative Change Assessment.

This volume seeks to provide inspiration and useful knowledge for practitioners, policymakers, and scientists to spearhead the debate on transformative change in various arenas. It will also provide a broader contribution to the knowledge base informing key policy processes, including those of IPBES and CBD. In doing so, it is our hope that the volume will help to advance the urgent action that is needed to address biodiversity loss and degradation, towards the vision of living in harmony with nature by 2050.

United Nations University Institute for   S. Yume Yamaguchi
the Advanced Study of Sustainability,
Tokyo, Japan

Preface

The Satoyama Initiative is "a global effort to realise societies in harmony with nature", started through a joint collaboration between the United Nations University (UNU) and the Ministry of the Environment of Japan. The initiative focuses on the revitalisation and sustainable management of "socio-ecological production landscapes and seascapes" (SEPLS), areas where production activities help maintain biodiversity and ecosystem services in various forms while sustainably supporting the livelihoods and well-being of local communities. In 2010, the International Partnership for the Satoyama Initiative (IPSI) was established to implement the concept of the Satoyama Initiative and promote various activities by enhancing awareness and creating synergies among those working with SEPLS. IPSI provides a unique platform for organisations to exchange views and experiences and to find partners for collaboration. At the time of writing, 271 members have joined the partnership, including governmental, intergovernmental, non-governmental, private sector, academic, and indigenous peoples' organisations.

The Satoyama Initiative promotes the concept of SEPLS through a three-fold approach that argues for connection of land- and seascapes holistically for management of SEPLS (see Fig.1). This often means involvement of several sectors at the landscape scale, under which it seeks to: (1) consolidate wisdom in securing diverse ecosystem services and values, (2) integrate traditional ecosystem knowledge and modern science, and (3) explore new forms of co-management systems. Furthermore, activities for SEPLS conservation cover multiple dimensions, such as equity, addressing poverty and deforestation, and incorporation of traditional knowledge for sustainable management practices in primary production processes such as agriculture, fisheries, and forestry (UNU-IAS & IGES 2015).

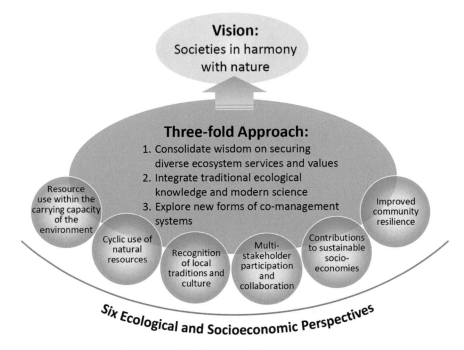

Fig. 1 The conceptual framework of the Satoyama Initiative

As one of its core functions, IPSI serves as a knowledge-sharing platform through the collection and sharing of information and experiences on SEPLS, providing a place for discussion among members and beyond. Close to 230 case studies have been collected and are shared on the IPSI website, providing a wide range of knowledge covering diverse issues related to SEPLS. Discussions have also been held to further strengthen IPSI's knowledge-facilitation functions, with members suggesting that efforts should be made to produce knowledge on specific issues in SEPLS in order to make more targeted contributions to decision-makers and on-the-ground practitioners.

It is in this context that a project to create a publication series titled the "Satoyama Initiative Thematic Review" (SITR) was initiated in 2015 as a joint collaboration between UNU's Institute for the Advanced Study of Sustainability (UNU-IAS), which hosts the IPSI Secretariat, and the Institute for Global Environmental Strategies (IGES), an IPSI partner and research institute based in Japan. The Thematic Review was developed as a compilation of case studies providing useful knowledge and lessons focusing on a specific theme that is important for SEPLS. The overall aim of the Thematic Review is to collect experiences and relevant knowledge, especially from practitioners working on the ground, considering their usefulness in providing concrete and practical knowledge and information as well as their potential to contribute to policy recommendations. Each volume is also accompanied by a synthesis chapter which extracts lessons learned through the case studies, presenting them for policy-relevant academic discussions. This series also contributes to efforts being made by researchers to strengthen the evidence base for policymaking concerning socio-ecological dynamics and resilience, including those under the Intergovernmental Science-Policy Platform on Biodiversity and Ecosystem Services (IPBES) and the Convention on Biological Diversity (CBD).

Five volumes of the SITR series have been published by UNU on an annual basis since 2015. The first volume with the theme "enhancing knowledge for better management of SEPLS" focused on ways to identify, collect, document, maintain, exchange, refine, augment, and make use of information and knowledge for better management of SEPLS. The second volume's theme was "mainstreaming concepts and approaches of SEPLS into policy and decision-making", covering topics including advocacy, multi-stakeholder engagement, facilitation and coordination of institutions, concrete tools, and information useful for policymakers and stakeholders. The third volume, titled "sustainable livelihoods in SEPLS" identified drivers linked to sustainable livelihoods in SEPLS that are crucial to meet needs for human well-being and to foster sustainable use of natural resources. The fourth volume, "sustainable use of biodiversity in SEPLS and its contribution to effective area-based conservation" looked at how effective management of SEPLS, which can include areas inside and outside of designated protected areas, can achieve benefits for both biodiversity conservation and human livelihoods through sustainable use of biodiversity. The fifth volume on "understanding the multiple values associated with sustainable use in SEPLS" examined intrinsic, instrumental, and relational values provided and maintained through sustainable use of biodiversity in SEPLS management.

This volume is the first in the SITR series to be published by Springer in an aim to reach out to a broader range of readers beyond our institutional territories while maintaining the volume's scope and relevance to science–policy–practice interfaces. Like previous volumes, this publication was developed through a multistage process including both peer review and discussion among the authors at a workshop. The authors had several opportunities to receive feedback, which helped them to improve their manuscripts in substance, quality, and relevance. First, each manuscript received comments from the editorial team relating primarily to their contributions to the theme of the volume. Peer review was then conducted by the authors of other chapters, with each author receiving feedback from two other authors who were requested to comment on whether the manuscript was easy to understand and informative, provided useful lessons, and so on. The aforementioned workshop was then held to enable the exchange of feedback between authors. The workshop was held virtually in June 2020 because of the travel restrictions due to the COVID-19 pandemic and was organised separately for two regional groups to accommodate time differences among participants, although past workshops for previous volumes were always held in person. Here, the authors presented their case studies and received comments both from the two designated reviewers and from the other workshop participants. The basic ideas contained in the synthesis of the concluding chapter were developed from the presentations and discussions during the workshop, and the chapter was made available for review by authors before finalisation.

Our experience producing these volumes leads us to believe that the above process offers an opportunity for authors from both academic and non-academic organisations to contribute to generating knowledge in an accessible and interactive way, as well as to provide high-quality papers written in simple language for academics and a broader audience alike. It is our hope that this publication will be useful in providing information and insights to practitioners, researchers, and policymakers on the importance of long-term management of SEPLS for facilitating transformative change to move towards a sustainable world. This, we hope, will prompt policymaking that strengthens such integrated and holistic management approaches.

We would like to thank all of the authors who contributed their case studies. We are also grateful to the other participants in the case study workshop who provided insightful remarks and valuable inputs into the discussions. These individuals include Eduardo S. Brondizio, Keiichi Nakazawa, Mahefatiana Ralisata, Eiji Tanaka, Chad Tudenggongbu, and Mari Yamazaki. Special thanks go to the co-editors from IGES for their continued collaboration in the publication process of this volume. We also thank other colleagues of UNU-IAS and IGES who were supportive and instrumental in organising the workshop and facilitating the publication process: William Dunbar, Saeko Kadoshima, Raffaela Kozar, Yoshino Nakahara, Miyuki Noguchi, Yasukuni Shibata, Akio Takemoto, Hiroaki Takiguchi, Nicholas Turner, Makiko Yanagiya, Evonne Yiu, and Kanako Yoshino. Our gratitude extends to the UNU administration, especially Francesco Foghett, Florence Lo, Daniel Powell, and to colleagues from UN Geospatial Information Section for all the instrumental support through the publication process. Furthermore, we acknowledge Susan Yoshimura who skillfully proofread the manuscripts.

Publication of this volume as a Springer Book would not have been possible without the generous support of Kazuhiko Takeuchi, Osamu Saito, and Joanne M. Kauffman. Our grateful thanks are also due to the Ministry of the Environment, Japan for supporting the activities of IPSI and its secretariat hosted by UNU-IAS.

Tokyo, Japan   Maiko Nishi
                                                              Suneetha M. Subramanian
                                                                       Himangana Gupta
                                                                         Madoka Yoshino

Reference

UNU-IAS and IGES (eds.) 2015, Enhancing knowledge for better management of socio-ecological production landscapes and seascapes (SEPLS) (Satoyama Initiative Thematic Review vol.1), United Nations University Institute for the Advanced Study of Sustainability, Tokyo.

Contents

Chapter 1
Introduction

Maiko Nishi, Suneetha M. Subramanian, Himangana Gupta,
Madoka Yoshino, Yasuo Takahashi, Koji Miwa, and Tomoko Takeda

Abstract This chapter introduces the idea of transformative change for sustainability and its relevance to the concept and practices of socio-ecological production landscapes and seascapes (SEPLS). First, it lays out the context where transformative change has been described as a way of fundamental, system-wide reorganisation of technological, economic and social factors to achieve the global goals of sustainability and nature conservation. Following a literature review, which offers the current state of knowledge concerning transformative change, the chapter discusses how SEPLS management relates to the idea of transformative change. In particular, it highlights the potentials of integrated approaches to managing SEPLS that can result in multiple benefits beyond biodiversity conservation and facilitate transformative change while addressing well-being needs and challenges specific to the local contexts. With this background and conceptual underpinning, the chapter provides the scope and objectives of the book as well as the key questions followed by the case study chapters. Finally, it introduces the organisation of the book and presents an overview of the case studies.

Keywords Socio-ecological production landscapes and seascapes · Transformative change · Sustainable pathways · Systemic change · Landscape approaches · Case studies · Science-policy-practice interface · Sustainable development · Biodiversity conservation

M. Nishi (✉) · H. Gupta
United Nations University Institute for the Advanced Study of Sustainability (UNU-IAS),
Tokyo, Japan
e-mail: nishi@unu.edu

S. M. Subramanian · M. Yoshino
United Nations University Institute for the Advanced Study of Sustainability (UNU-IAS),
Tokyo, Japan

Institute for Global Environmental Strategies (IGES), Hayama, Japan

Y. Takahashi · K. Miwa · T. Takeda
Institute for Global Environmental Strategies (IGES), Hayama, Japan

1.1   What Do We Know About Transformative Change
for Sustainability?

The idea of "transformative change" has been gaining more attention as something
that is needed to deal with today's environmental and developmental problems. The
2030 Agenda for Sustainable Development advocates taking "the bold and transformative steps which are urgently needed to shift the world on to a sustainable and
resilient path" (UN 2015, p. 3). The Intergovernmental Panel on Climate Change
(IPCC) Special Report on the impacts of global warming of 1.5 °C calls for
"transformative systemic change, integrated with sustainable development" (IPCC
2018, p. 40). The Intergovernmental Science-Policy Platform on Biodiversity and
Ecosystem Services (IPBES) Global Assessment Report, launched in May 2019,
also cautions that goals for conserving and sustainably using nature and achieving
sustainability, including the 2030 Agenda for Sustainable Development and the
2050 Vision for Biodiversity, cannot be met by ongoing trajectories. It thus urges
"transformative changes across economic, social, political and technological factors"
in order to achieve these goals for 2030 and beyond (IPBES 2019a, p. 33).

But what does "transformative change" mean? According to IPBES, it refers to
"[a] fundamental, system-wide reorganization across technological, economic and
social factors, including paradigms, goals and values" (IPBES 2019a, p. 14). The
IPBES global assessment conceptualises the governance of transformative change as
shown in Fig. 1.1. Direct drivers, including changes in land and sea use, direct
exploitation, climate change, pollution, and invasion of alien species, are the result
of indirect drivers such as demographic and sociocultural factors, economic and
technological aspects, institutions and governance, disasters, conflicts, and epi-
demics. Both direct and indirect drivers have been accelerated over the past
50 years. Five main interventions, or levers, are proposed to generate transformative
change: (1) incentives and capacity building; (2) cross-sectoral cooperation;
(3) pre-emptive action; (4) decision-making in the context of resilience and uncertainty; and (5) environmental law and implementation. Also, eight priority points of
intervention have been found as leverage points that are likely to yield large impacts:
(1) visions of a good life; (2) total consumption and waste; (3) values and action;
(4) inequalities; (5) justice and inclusion in conservation; (6) externalities and
telecouplings; (7) technology, innovation and investment; and (8) education and
knowledge generation and sharing.

This concept of transformative change builds on a synthesis of diverse strands of
the literature, in which systemic change is a common subject. As described above,
the notion of systemic change is also implicitly expressed in the synonymous terms
used by other international processes such as IPCC (e.g. transformative systemic
change). In the literature, the process of systemic change is, for instance,
characterised as a complex web of fast and slow developments cumulatively
resulting from positive and negative feedback mechanisms (Edmondson et al.
2019; Grin et al. 2010). In a successful transition, a new system adapts to the
changed internal and external circumstances and arrives at a higher degree of

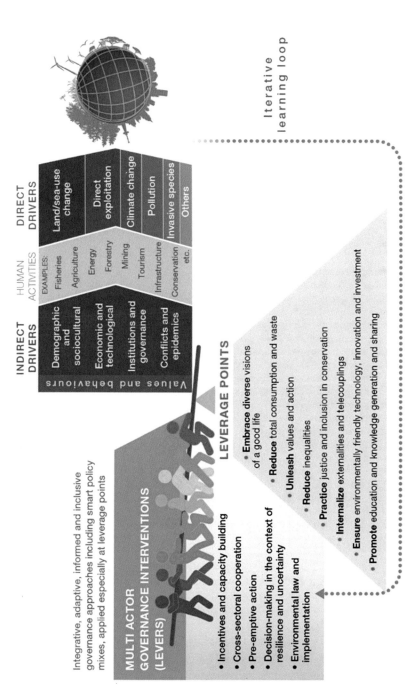

Fig. 1.1 Transformative change in global sustainability pathways (IPBES 2019a)

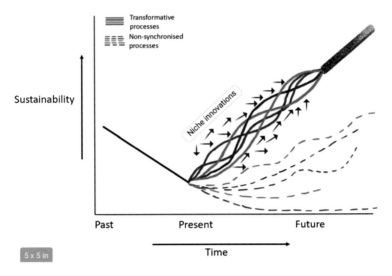

Fig. 1.2 Conceptual schematic of transformative change. Adapted from IPBES (2019a), Grin et al. (2010), Rotmans and Loorbach (2009) and Geels (2005)
Note: Over time (X axis), the level of sustainability (Y axis) has declined as exemplified by the trends related to biodiversity and other sustainability indices (IPBES 2019a). The figure depicts two broad potential scenarios in achieving sustainability goals moving from now towards the future—let us say 2050–, depending on the pathways adopted: (1) the pathways of transformative processes, and (2) the pathways of non-synchronised processes. In the transformative processes, each of the sectoral pathways (colored, solid lines, exemplifying different pathways in various sectors such as agriculture, tourism, fisheries, etc.) will undergo internal transformation involving positive and negative feedback mechanisms shown as non-linear and dynamic pathways (fluctuated) but will overall move towards higher levels of sustainability (moving upward remarkably). At the same time, these different pathways will influence and bolster each other to become harmoniously synchronised, whereas emerging niche agents stimulate niche development at the micro-level and cumulatively lead to niche innovations. Overall, these internal and external developments lead to bringing about transformative change (rainbow-colored, bold line). In the non-synchronised processes, some of the pathways (colored, dotted lines) will go through internal transformation (fluctuated), but others will not reflect positive and negative feedbacks much in their own systems (less fluctuated). While achieving varied levels of sustainability, the different interventions will be undertaken separately without synergistic effects and thus these pathways will not intersect and interact with each other in a synchronised manner, failing to bring about transformative change

complexity encompassing various organisational levels and multiple sectors (Rotmans and Loorbach 2009). For instance, a policy mix drives socio-technical change through multiple policy effects while leading to further policy mix advancement by influencing the policy processes along with the resultant feedback mechanisms (Edmondson et al. 2019).

Figure 1.2 describes a concept of this feedback process as possible pathways from present to future. In the transformative processes, positive and negative feedbacks inform and influence each other in a synchronised manner, strengthening the system in itself. Conversely, the non-synchronised processes involve a weaker feedback mechanism, leading to a limited level of sustainability or even ending up with a collapse of the system earlier. Taking food policy as an exemplary intervention, the food production system may be improved towards sustainability through various changes and actions including organic farming practices, entrepreneur initiatives, and consumers' reactions and feedbacks, but if it lacks synergy and consonance between different interventions, it might not reach a high level of sustainability, for instance, as expected to achieve the 2050 Vision for Biodiversity. Finally, adapting to internal and external changes that interact with each other—including policy innovations across different sectors, technological advancement, broad socio-economic changes, and environmental changes—would reinforce the system in question along with the development of other interventions, resulting in a new system that has arrived at a higher order of complexity involving multiple social organisations across different levels. For instance, this could be manifested as mainstreaming of biodiversity into agriculture, forestry, fisheries, tourism, business and industries, and many other sectors at multiple levels. The transformative processes thus enable achievement of ambitious global goals and targets.

Current status quo systems often inhibit sustainable development and sometimes constitute or exacerbate indirect drivers of biodiversity loss (IPBES 2019a). When an incumbent system reaches equilibrium, various factors impede internal forces to change the system (Grin et al. 2010). For instance, the food production system may not entail dynamic feedback mechanisms within the system or may not interact with other policy sectors to move towards sustainability. It is often the case that policymaking is pursued within one sector only at the national level, whereas no feedback mechanism is devised to incorporate inputs into and provide feedbacks from other sectors and levels. This results in not only a failure in bringing about transformative change, but sometimes leads the society to a further unsustainable future.

Nevertheless, newcomers or small groups of emerging niche agents (i.e. agents of change giving rise to niche innovations, such as environmental champions, local communities, and other non-state actors featured in the following case study chapters), who are yet to be absorbed into the incumbent equilibrium of systems, have the potential to break through conventional systems and establish new regimes with a set of dynamics that are better adapted to the changed environment or circumstances (Rotmans and Loorbach 2009). This multilateral process involves a multi-level perspective (MLP) which specifically attends to the interactions between macro-, meso- and micro-levels in system transition (Geels 2005). At the micro-level, a wide array of innovations emerge in niches and are gradually aligned and linked together to form a new configuration of social and technological elements. A change in the socio-technical landscape at the macro-level causes a misfit of a system pertinent to the existing regime, and thereby opens up an opportunity for a regime shift. A system transformation occurs when a new socio-technical configuration that addresses the system misfit links up to (or 'anchors') and penetrates throughout the regime.

To bring about transformative change, 'intermediaries' play a key role in facilitating collaboration among diverse stakeholders (Klerkx et al. 2012). In the final process of system transformation, 'scaling' of innovation occurs. Moore et al. (2015) postulate the notion of scaling out, up, and deep. Scaling-out is an attempt to impact greater numbers through deliberate replication and spreading principles; scaling-up is an effort to change laws, rules and policy; and scaling-deep is intended to change mindsets. Moore et al. (2015) highlight the need to integrate these three approaches, rather than focusing on one among others, to induce system-wide change.

As a means of inducing change, the notion of leverage points has developed, originating from the Meadows' theory (Meadows 1999), which describes 'shallow' and 'deep' leverage points to induce system-wide transformative change. Shallow leverage points are tangible and thus often the subject of policy interventions, but not strong enough to yield a system-wide transformation (Abson et al. 2017). These include parameters pertaining to production, flow and stock of substances and their structure, as well as the feedback mechanisms that are in place to regulate these parameters. Deep leverage points are mostly intangible and hard to alter but can bring about extraordinary impact once effectively addressed. These include system design, i.e. the structure of information flows, the rule of the system and the power to change or create system structure, as well as the intent of the system, including the goals of the system and the paradigm which the system serves. Abson et al. (2017) identified three realms that exert effects on these deep leverage points and thereby effectively contribute to sustainability transformation: institutions, people's connections to nature, and the production and use of knowledge in transformative change processes.

The IPBES global assessment illustrates plausible global pathways to sustainability, which are coherent with known constraints on economics, resource use and human development goals, but require transformative change as fundamental changes in development paradigms (IPBES 2019b). While exploring a (more) sustainable future that may unfold in a context-dependent and evolutionary manner with emergent properties (rather than in a deterministic and linear way), it highlights that implementation of instruments through integrative, informed, inclusive and adaptive place-based governance interventions can enable global transformation (IPBES 2019c). Furthermore, it points to a diversity of actors at multiple leverage points—ranging from intergovernmental organisations, governments, non-governmental organisations, and indigenous peoples and local communities to the private sector—who can apply the levers. Even after the launch of the IPBES global assessment, however, many assessment authors repeatedly received a question about "[w]hat does transformative change mean, and how do we get started?"—most probably due to the insufficient recognition or understanding of the links between theories and practices for bringing about transformative change (Chan 2019).

In view of the need for a better understanding of how transformative change can be brought about to inform development of policies and actions, IPBES decided in early 2019 to conduct a thematic assessment of the underlying causes of biodiversity loss and the determinants of transformative change and options for achieving the 2050 Vision for Biodiversity, the so-called "assessment on transformative change" as part of its rolling work programme (IPBES 2019d). Considering broader social and economic goals in the context of sustainable development, this assessment is aimed at understanding and identifying factors in human society at both the individual and collective levels—including behavioural, social, cultural, economic, institutional, technical and technological dimensions—which can be leveraged to bring about transformative change for the conservation, restoration and wise use of biodiversity (IPBES 2019b). The three-year assessment work will be launched upon the eighth session of the IPBES Plenary to be held in the near future. A recent review of the studies on sustainability transformation also identified a lack of empirical knowledge on its real-world examples (Salomaa and Juhola 2020). As such, knowledge concerning transformative change in connection to biodiversity and ecosystem services is yet to be assessed globally in a comprehensive manner, particularly with reference to practical examples.

1.2 Potential Contributions of Socio-Ecological Production Landscapes and Seascapes to Transformative Change

As a system-wide reorganisation that is needed for humanity to achieve global goals related to nature, transformative change requires consideration of the relationships and linkages between SDGs, targets towards the 2050 Vision for Biodiversity, and the Paris Agreement on climate change, and between related conventions like the Convention on Biological Diversity (CBD), United Nations Framework Convention on Climate Change (UNFCCC) and United Nations Convention to Combat Desertification (UNCCD). Interlinkages exist inherently between different global environmental problems across scales and levels given the complex interdependency of food, water, and energy among competing uses, which are further compounded by climate change (Rasul and Sharma 2016). While the nexus among multiple ecosystem services is gaining prominence as a methodological approach to resource management so as to address sustainability challenges and improve policymaking, these problems cannot be resolved without local actions (Cremades et al. 2019; Rasul and Sharma 2016).

The Satoyama Initiative promotes integrated approaches with a focus on the locally or regionally based revitalisation and management of socio-ecological production landscapes and seascapes (SEPLS). Being portrayed as a mosaic of various types of ecosystems (e.g. farmlands, secondary forests, wetlands, coastal zones and human settlements), SEPLS refer to the areas where production activities help maintain biodiversity and ecosystem services in various forms while sustainably supporting the livelihoods and well-being of local communities. Nature provides multiple benefits for people (e.g. material goods and spiritual inspiration) through biophysical processes and ecological interactions with anthropogenic assets (e.g. knowledge, infrastructure, technology and institutions) not only at the local level but across a wide range of communities (IPBES 2019a).

However, multiple human drivers, including both direct and indirect ones, have increasingly and significantly altered nature (e.g. land surface, ocean and wetlands) during the past 50 years, accelerating the rate of species extinction and devastating global ecosystems (IPBES 2019a). This in turn threatens a good quality of life through the degradation of nature's contributions to people and undermines efforts to achieve many of the international societal and environmental goals (IPBES 2019a). It is important to note that SEPLS, which are purposively managed to produce multiple ecosystem services, contribute directly to the well-being of local communities but also to that of a larger population outside their boundaries, thereby supporting local, national and global economies (Gu and Subramanian 2014). At the same time, the production processes within SEPLS are increasingly subject to external demands and pressures and influenced by policy decisions at the national and international levels (Gu and Subramanian 2014).

Integrated approaches to managing SEPLS can result in multiple benefits beyond biodiversity conservation, including provision of ecosystem services, preservation of traditional knowledge and practices, climate change mitigation and adaptation, ecosystem restoration, and social equity and rights. SEPLS management manifests integrated approaches on a landscape or seascape scale (often called landscape approaches), which offer opportunities to reconcile multiple interests, values, and forms of resource use. In particular, these approaches help deliberate sustainable pathways by bringing together diverse stakeholders operating on the landscape or seascape, specifically recognising trade-offs and power asymmetries among them (IPBES 2019a; Sayer et al. 2017).

As small groups of niche agents have the potential to make a breakthrough for transformative change (Grin et al. 2010), good practices at the local level are critical to achieve global goals. In fact, as demonstrated in the previous volumes of the Satoyama Initiative Thematic Review (SITR) from 2015 to 2019, the Satoyama Initiative involves many case studies showing how these approaches contribute to global goals through local actions by bringing together all the different concerns and interests in the landscape or seascape. These cases can be seen as real-world examples of transformative change or the seeds for it. SEPLS management could thus provide practical and experience-based insights for understanding and gauging transformative change and identifying determinants of such change. Furthermore, multi-level networks such as the International Partnership for the Satoyama Initiative (IPSI), which can link an array of locally-relevant solutions across ecosystems and scales, help promote new actions and policy in response to challenges and opportunities to achieve biodiversity conservation, ecosystem restoration and more broadly sustainable development (Kozar et al. 2019). Given that transformative change has been called for in policymaking and implementation processes as mentioned above, exploring the contributions of SEPLS to transformative change would also have strong policy significance for the achievement of relevant global goals.

1.3 Objectives and Structure of the Book

The primary focus of this book is the relevance of SEPLS to transformative change. The book aims to provide insights on how SEPLS management on the ground can contribute to more sustainable management and achievement of global goals for sustainable development through bringing about transformative change. Considering integrated approaches to SEPLS management can deliver multiple benefits for people and the planet, this volume brings together case studies on SEPLS management from different regions around the world, which delve into the relevance of SEPLS to various aspects of transformative change. The case studies highlight the roles, attitudes and actions of those responsible for management, including smallholders, indigenous peoples and local communities, and other stakeholders in conserving biodiversity while ensuring that SEPLS provide other benefits (e.g. food security, water quality, health, quality of life, enhanced carbon storage, reduced footprint of cities). Furthermore, they attend to how SEPLS management may have implications for national and global policymaking processes.

In particular, the case studies address the following questions:

- How has SEPLS management helped in pursuing transformative change or leading to the emergence (i.e. seeds) of transformative change?
- What indicators and methods are used to assess the achievements for transformative change?
- What are the roles, attitudes and actions of those responsible for management, including smallholders, indigenous peoples and local communities, in facilitating transformative change while ensuring the multiple benefits from SEPLS? In this regard, are there any policy implications at the local, regional, national and/or global levels?
- What are the values underpinning SEPLS management and how do they contribute to bringing about transformative change for improved sustainability?
- What are the challenges and opportunities in bringing about transformative change towards a sustainable world through SEPLS management?

The following Chaps. 2, 3, 4, 5, 6, 7, 8, 9, 10, 11, and 12 present eleven case studies from different parts of the world, including two from Africa, four from Asia, three from Europe, and two from Latin America (Fig. 1.3). Although each of the case studies features SEPLS that encompass different types of ecosystems, the case studies are largely grouped into dominant landscapes or seascapes as follows: (1) mountain landscapes (Chaps. 2, 3, 4, and 5); (2) agricultural landscapes (Chap. 6); (3) watershed landscapes (Chaps. 7, 8, 9, and 10); and (4) coastal landscapes or seascapes (Chaps. 11 and 12) (Table 1.1). Most of the cases primarily focus on the efforts of SEPLS management at the local level, including those in one or multiple local communities (Chaps. 2, 3, 4, 5, 6, 10, and 11) or a certain catchment or a bay area (Chaps. 7, 9 and 12), while some of them have scaled up to national-scale initiatives or have been replicated in other regions. One exception is a cross-scale comparison between local and regional initiatives (Chap. 8). The time scale of

Fig. 1.3 Locations of the case studies by types of landscapes (Map template: Geospatial Information Section, United Nations) Note: Details of the case study locations, including geographic coordinates, are described in each chapter

Table 1.1 Overview of the case studies

Focused landscape or seascape	Chapter (Country)	Location and ecosystem type	Spatial scale or jurisdictional level[a]	Temporal scale[b]	Problems	Objectives
Mountain landscapes	Chapter 2 (Kenya)	Mountain forest, farmland	Forest fragments (Taita Hills landscape, 5 forests surrounded by villages: 769.8 ha)	2012–2019	• Environmental: forest degradation and its adverse effect on water resources, habitat reduction, loss of flora and fauna diversity, soil erosion • Social: poverty, food insecurity, societal conflicts	Forest restoration, livelihood improvement, participatory management, climate change mitigation and adaptation
	Chapter 3 (Madagascar)	Mountain forest, farmland	Village (principal village Ambondro and 5 other villages: 4680 ha)	2013–2020	• Environmental: deforestation, biodiversity loss, habitat reduction • Social: food insecurity, poverty, population growth (insufficiency of food production, shortage of raw materials) • Political: corruption	Agroforestry landscaping, local reforestation with indigenous species, forest restoration, food production, livelihood improvement
	Chapter 4 (India)	Mountain forest, farmland, pasture	Village (Khaljuni village: including 68 ha farms and 436 ha forest)	1991–2016	• Environmental: farmland abandonment, forest degradation	Forest restoration, biodiversity conservation, climate change mitigation
	Chapter 5 (Nepal)	Mountain forest, farmland	Four villages (in four municipalities in Gandaki Province)	2015–2019	• Environmental: landslides, climate change-related hazards, farmland abandonment • Social: food insecurity,	Livelihood improvement, improved ecosystem services (water availability and retention, food production),

(continued)

Table 1.1 (continued)

Focused landscape or seascape	Chapter (Country)	Location and ecosystem type	Spatial scale or jurisdictional level[a]	Temporal scale[b]	Problems	Objectives
					financial stress, rural youth outmigration	biodiversity conservation
Agricultural Landscapes	Chapter 6 (Italy)	Farmland (hilly terrain)	Town (Montespertoli in Florence metropolitan area: approx. 12,000 ha)	2013–2020 (project started in 2014)	• Environmental: farmland abandonment, unsustainable agriculture, loss of species diversity • Social: decline in agricultural activities (e.g. town fairs, festivals), decrease in farming population	Sustainable agriculture, reduced carbon footprint, reduced landslides and waste, biodiversity conservation, livelihood improvement (farmers' revenues), improved health
River-basin/ watershed Landscapes	Chapter 7 (Chinese Taipei)	Upstream watershed, farmland (hillside, slope land)	Upstream community (Han River upstream watershed: 250 ha containing 100 ha farmlands and 80 ha foothills)	2012–2019	• Environmental: low water retention capacity, uneven rainfall pattern, water shortage (overuse of groundwater) • Social: aging labour force, rural youth outmigration, water-use conflicts	Landscape restoration, sustainable local economy, improved water management, biodiversity conservation, community empowerment
	Chapter 8 (Spain)	Peri-urban watershed, forest (peri-urban), riparian and coastal zones (peri-urban)	• Regional: autonomous community of Galicia (2958 ha) • Local: neighbourhood association of Chapela in Galicia (975 ha)	2017–2019	• Environmental: pollution, increased ecosystem vulnerability (riparian and forest) • Social: decreased access to green infrastructure, mobility limitations	Green infrastructure development, biodiversity conservation, pollution reduction, improved air quality, improved hydrological regulations, landscape quality,

				Period[b]	Issues	Outcomes
					(access to public transportation and pedestrian mobility)	sense of place, cultural heritage linked to ecosystems
	Chapter 9 (United Kingdom)	Upstream watershed, forest, farmland, pasture, wetlands	Upstream catchment (Upper Thames catchment: 26,000 ha including 19 parish councils and 22,000 ha of farmland)	2013–2016	• Environmental: water pollution, flooding	Improvement of water quality, community empowerment
	Chapter 10 (Colombia)	Downstream watershed, lowland forest, riparian and coastal zones	Ethnic community (San Marcos: 3689 ha)	2013–2020	• Environmental: ecosystem degradation (woodland exploitation, mining, dam construction) • Social: poverty, loss of cultural identity, lack of education, social conflicts (illegally-armed groups, drug trafficking)	Biodiversity conservation, sustainable natural resource use
Coastal landscapes or seascapes	Chapter 11 (Philippines)	Coast (rural), mangroves, downstream watershed	Village (Alitas, an estuarine barangay: 676 ha, including 355 ha of mangrove forest)	2013–2020	• Environmental: climate change-related hazards • Social: impacts on coastal livelihoods and well-being	Enhanced local resilience to climate change, mangrove conservation
	Chapter 12 (Antigua and Barbuda)	Coast (urban), marine, island, watershed (urban, peri-urban, and rural)	Bay flashes (Hanson's Bay Flashes—Key Biodiversity Area in St. John's City: 20 ha)	2018–2020	• Environmental: water and marine pollution, solid waste mistreatment, risks to human health • Political: lack of policy for waste control	Waste reduction and recycling, awareness raising and capacity development for recycling, livelihood improvement

[a]Refers to study sites or an area where the initiative of SEPLS management was primarily conducted, although some diffusive effect on surrounding areas or replication in other regions was observed in several cases (e.g. Chaps. 5, 6 and 12)
[b]Refers to the period of time during which changes were observed and analysed for each case study

the changes observed in the case studies ranges from the last few years (Chaps. 8, 9, and 12) and 5–10 years (Chaps. 2, 3, 5, 6, 7, 10, and 11) to more than two decades (Chap. 4). All the cases illustrate unique initiatives to address particular environmental problems (e.g. ecosystem degradation, habitat loss, and pollution) but often in combination with social problems (e.g. poverty, food insecurity, demographic decline and social conflicts), whereas political problems (e.g. corruption, lack of control) sometimes inhibit resolution or exacerbate the social and ecological problems. Importantly, many of the cases exemplify initiatives not only to address the immediate problems but to collectively identify long-term solutions and ensure continuous delivery of multiple benefits from SEPLS.

With the general understanding of transformative change (i.e. a fundamental, system-wide reorganisation as defined by IPBES), the case studies commonly address the above key questions to elucidate the relevance of each SEPLS management to aspects of transformative change. As a concluding chapter, Chap. 13 synthesises key findings from the case studies and draws out key messages to offer implications for science, policy and practice as well as their interfaces in moving towards a sustainable world. By revisiting the existing conceptual frameworks described in this chapter, the last chapter re-examines the concept of transformative change. As discussed in Chap. 13, most of the case studies demonstrate seeds of change that have great potential to facilitate and pursue sustainable transformation, while highlighting challenges and opportunities to bring about transformative change as a groundbreaking system-wide transformation. Despite the limitations in terms of extent, scope and depth of change, the case studies offer critical insights to elaborate the concept of transformative change and advance methodologies for monitoring and evaluation on progress in pursuing transformative change.

References

Abson, D. J., Fischer, J., Leventon, J., Newig, J., Schomerus, T., Vilsmaier, U., von Wehrden, H., Abernethy, P., Ives, C. D., Jager, N. W., & Lang, D. J. (2017). Leverage points for sustainability transformation. *Ambio, 46*, 30–39.

Chan, K. (2019). What is transformative change, and how do we achieve it? *IPBES News*. Retrieved 28 July, 2020, from https://ipbes.net/news/what-transformative-change-how-do-we-achieve-it

Cremades, R., Mitter, H., Tudose, N. C., Sanchez-Plaza, A., Graves, A., Broekman, A., Bender, S., Giupponi, C., Koundouri, P., Bahri, M., Cheval, S., Cortekar, J., Moreno, Y., Melo, O., Karner, K., Ungurean, C., Davidescu, S. O., Kropf, B., Brouwer, F., & Marin, M. (2019). Ten principles to integrate the water-energy-land nexus with climate services for co-producing local and regional integrated assessments. *Science of The Total Environment, 693*, 133662.

Edmondson, D. L., Kern, F., & Rogge, K. S. (2019). The co-evolution of policy mixes and socio-technical systems: Towards a conceptual framework of policy mix feedback in sustainability transitions. *Research Policy, 48*(10).

Geels, F. W. (2005). Processes and patterns in transitions and system innovations: Refining the co-evolutionary multi-level perspective. *Technological Forecasting and Social Change, 72*, 681–696.

Grin, J., Rotmans, J., & Schot, J. (2010). Transitions to sustainable development: New directions in the study of long term transformative change.

Gu, H., & Subramanian, S. M. (2014). Drivers of change in socio-ecological production landscapes: Implications for better management. *Ecology and Society, 19.*

IPBES. (2019a). *Summary for policymakers of the global assessment report on biodiversity and ecosystem services of the intergovernmental sciencepolicy platform on biodiversity and ecosystem services* (pp. 56). In S. Díaz, J. Settele, E. S. Brondízio, H. T. Ngo, M. Guèze, J. Agard, A. Arneth, P. Balvanera, K. A. Brauman, S. H. M. Butchart, K. M. A. Chan, L. A. Garibaldi, K. Ichii, J. Liu, S. M. Subramanian, G. F. Midgley, P. Miloslavich, Z. Molnár, D. Obura, A. Pfaff, S. Polasky, A. Purvis, J. Razzaque, B. Reyers, R. Roy Chowdhury, Y. J. Shin, I. J. Visseren-Hamakers, K. J. Willis, & C. N. Zayas (Eds.). Bonn, Germany: IPBES Secretariat.

IPBES. (2019b). Next work programme of the Platform, IPBES/7/6.

IPBES. (2019c). *Global assessment report on biodiversity and ecosystem services of the Intergovernmental Science-Policy Platform on Biodiversity and Ecosystem Services.* Bonn, Germany: IPBES Secretariat.

IPBES. (2019d). Rolling work programme of the Intergovernmental Science-Policy Platform on Biodiversity and Ecosystem Services up to 2030, Decision IPBES-7/1.

IPCC. (2018). Global Warming of 1.5°C: An IPCC Special Report on the impacts of global warming of 1.5°C above pre-industrial levels and related global greenhouse gas emission pathways, in the context of strengthening the global response to the threat of climate change, sustainable development, and efforts to eradicate poverty. In V. Masson-Delmotte, O. Zhai, H.-O. Pörtner, D. Roberts, J. Skea, P. R. Shukla, A. Pirani, W. Moufouma-Okia, C. Péan, R. Pidcock, S. Connors, J. B. R. Matthews, Y. Chen, X. Zhou, M. I. Gomis, E. Lonnoy, T. Maycock, M. Tignor, & T. Waterfield (Eds.). https://www.ipcc.ch/site/assets/uploads/sites/2/2019/05/SR15_Citation.pdf

Klerkx, L., van Mierlo, B., & Leeuwis, C. (2012). Evolution of systems approaches to agricultural innovation: Concepts, analysis and interventions. In I. Darnhofer, D. Gibbon, & B. Dedieu (Eds.), *Farming systems research into the 21st century: The new dynamic* (pp. 457–483). Dordrecht: Springer.

Kozar, R., Galang, E., Alip, A., Sedhain, J., Subramanian, S., & Saito, O. (2019). Multi-level networks for sustainability solutions: The case of the International Partnership for the Satoyama Initiative. *Current Opinion in Environmental Sustainability, 39,* 123–134.

Meadows, M. (1999). *Leverage points: places to intervene in a system.* Hartland, VT: The Sustainability Institute.

Moore, M.-L., Riddell, D., & Vocisano, D. (2015). Scaling out, scaling up, scaling deep strategies of non-profits in advancing systemic social innovation. *Journal of Corporate Citizenship, 2015,* 67–84.

Rasul, G., & Sharma, B. (2016). The nexus approach to water–energy–food security: An option for adaptation to climate change. *Climate Policy, 16,* 682–702.

Rotmans, J., & Loorbach, D. (2009). Complexity and transition management. *Journal of Industrial Ecology, 13,* 184–196.

Salomaa, A., & Juhola, S. (2020). How to assess sustainability transformations: A review. *Global Sustainability, 3,* e24.

Sayer, J. A., Margules, C., Boedhihartono, A. K., Sunderland, T., Langston, J. D., Reed, J., Riggs, R., Buck, L. E., Campbell, B. M., Kusters, K., Elliott, C., Minang, P. A., Dale, A., Purnomo, H., Stevenson, J. R., Gunarso, P., & Purnomo, A. (2017). Measuring the effectiveness of landscape approaches to conservation and development. *Sustainability Science, 12,* 465–476.

UN. (2015). *Transforming our world: The 2030 agenda for sustainable development, A/RES/70/1.* New York, NY: UN.

The opinions expressed in this chapter are those of the author(s) and do not necessarily reflect the views of UNU-IAS, its Board of Directors, or the countries they represent.

Chapter 2
Reconciling Community Livelihood Needs and Biodiversity Conservation in Taita Hills Forests for Improved Livelihoods and Transformational Management of the Landscape

Chemuku Wekesa, Leila Ndalilo, and Carolyne Manya

Abstract The fragmented forests of Taita Hills form an exceptional multi-functional socio-ecological production landscape with outstanding diversity of flora and fauna that provide ecosystem goods and services supporting human wellbeing and livelihood systems. However, these forests are threatened by illegal logging for wood products and encroachment for crop farming. A study was conducted in villages surrounding five forest fragments to establish the conservation programmes responsible for keeping these forests intact for provision of goods and services to the local communities. Semi-structured questionnaires were used to collect data from 250 respondents in 25 villages surrounding the five forest fragments. Twenty-five focus group discussions (FGDs) were held with key informants actively involved in conservation activities. Results showed that the Taita community conserves the forest fragments through management practices that integrate livelihood needs in conservation, such as butterfly farming, bee-keeping and eco-tourism. Additionally, community tree nurseries have been established to produce seedlings for restoring degraded areas, and agroforestry belts have been established on the forests' edges to provide wood products and protect the forests from encroachment. Likewise, village committees have been established to oversee conservation activities inside the village jurisdictional area. The integrated conservation and livelihood approach has reduced forest destruction, enhanced landscape connectivity for biodiversity conservation, increased incomes, enhanced capacity of the community to adapt to climate change, improved food security, enhanced carbon storage, strengthened traditional knowledge and practices, and ensured availability of clean water for the local population.

C. Wekesa (✉) · C. Manya
Kenya Forestry Research Institute, Taita Taveta Research Centre, Wundanyi, Kenya

L. Ndalilo
Kenya Forestry Research Institute, Coast Eco-Region Research Programme, Malindi, Kenya

Keywords Biodiversity · Community livelihoods · Landscape · Participatory forest management · Taita hills forests

2.1 Introduction

The Taita Hills forests form the northeastern part of the Eastern Arc Mountains, a mountain range with an exceptionally high degree of endemism and conservation value (Myers et al. 2000; Burgess et al. 2007; Hall et al. 2009). The forests are among 34 areas around the world considered global biodiversity hotspots (Omoro et al. 2010). The overriding conservation challenges in the Taita Hills forests are loss, fragmentation and degradation of indigenous forest cover. The indigenous forests have declined and become fragmented and degraded as a result of deforestation and planting of exotic tree species in degraded sites formerly part of indigenous forest (Rogo and Oguge 2000; Pellikka et al. 2009).

Currently, indigenous forests only cover about 430 ha, reflecting a 98% forest reduction over the last 200 years (Adriaensen et al. 2006). A recent study indicates that the deforestation rate stands at 0.5% annually (Wekesa 2018). The habitat reduction has led to increased isolation of the remaining forest patches, increased edge effects, soil erosion and negative hydrological effects. Likewise, the isolated forest fragments are embedded in the agricultural landscape, hence ranked among the most threatened biodiversity hotspots globally (Newmark 1998; Pellikka et al. 2009). Low to high levels of disturbance have been reported in the forest fragments. The forests of Mbololo and Ngangao, Chawia and Fururu, and Vuria have low, medium and high disturbance levels, respectively (Wekesa et al. 2016). The disturbances have altered the forest structure and reduced tree species diversity (Wekesa et al. 2016). Edge effects due to fragmentation of the forests have significantly affected the species diversity, distribution and abundance (Wekesa et al. 2018, 2019). Soil conditions such as moisture, nitrogen and potassium have also been affected by edge effects (Wekesa et al. 2018). Due to the small and highly isolated nature of forest fragments in Taita Hills, edge effects, and ongoing anthropogenic disturbances in these forests, a large proportion of the tree species is highly threatened, and is of immediate conservation concern. The occurrence of endemic species such as *Coffea fadenii* and *Afrocarpus usambarensis* is restricted to a narrow range, and their distribution is highly fragmented due to the fragmented nature of Taita Hills forests, threatening the two species with extinction. Hence, there is an urgent need to develop effective long-term conservation strategies that can save these species and other woody species growing along with them, namely *Macaranga conglomerata*, *Meineckia ovata*, *Memecylon teitense*, *Millettia oblata*, *Ocotea usambarensis*, *Psychotria petitii*, *Psychotria crassipetala* and *Prunus africana*.

Promotion of an integrated landscape management approach is urgently needed to sustainably manage the fragmented forests of Taita Hills and associated agricultural landscape in order to enhance ecological connectivity, conserve biodiversity and support local economies and livelihoods. This case study highlights practical and innovative community-led conservation activities that are reconciling community livelihood needs and biodiversity conservation. These activities have resulted in transformative change, including reduced destruction of forests, enhanced connectivity of forest fragments to conserve the remaining biodiversity, increased income for the local population, enhanced capacity of the community to adapt to climate change, improved food security, enhanced carbon storage, strengthened traditional knowledge and cultural practices and ensured availability of clean water from the forests that serve as key catchment areas.

2.2 Materials and Methods

2.2.1 Study Sites and Communities

The study was undertaken in Taita Taveta County, mainly inhabited by the Taita, Taveta and Kamba communities, and included villages surrounding the five main forest fragments in Taita Hills, which exhibit a wide range of conditions (Fig. 2.1 and Table 2.1). These forest fragments are Mbololo, Ngangao, Chawia, Fururu and Vuria. The forest fragments are in close proximity to a densely populated agricultural landscape. The sites were selected because of their rich biodiversity and the diverse agro-ecosystems in the surrounding farmlands. Indigenous forests are being lost to encroachment for crop farming, resulting in remarkable changes in indigenous forest areas and declining populations of endemic flora and fauna species (Pellikka et al. 2009). Natural resource use and management practices are guided by the traditional 'mitengo'[1] management system administered by the village conservation committees, which restricts activities that impact negatively on biodiversity.

About 52.7% of the population lives below the poverty line (CIDP 2013). The rainfall pattern is bimodal. A long rainy season occurs from March to May, with a short rainy season between November and December, but mist and cloud precipitation is a year-round phenomenon. The annual average rainfall is 1132 mm, with yearly maximum rainfall of about 2000 mm. The average temperature in Taita Taveta County is 23 °C with variations between 18 °C and 24.6 °C, with the hills experiencing lower temperatures of 18.2 °C compared to the lower zones with an average temperature of 24.6 °C (CIDP 2013). The relative humidity ranges from 79% to 83% (CIDP 2013).

---

[1]Small-sized community forests occurring in critical catchment areas and rich in endemic flora and fauna that are managed by community conservation committees for provision of ecosystem services, mainly water and biodiversity conservation, and are used as cultural welfare centers for the Taita community.

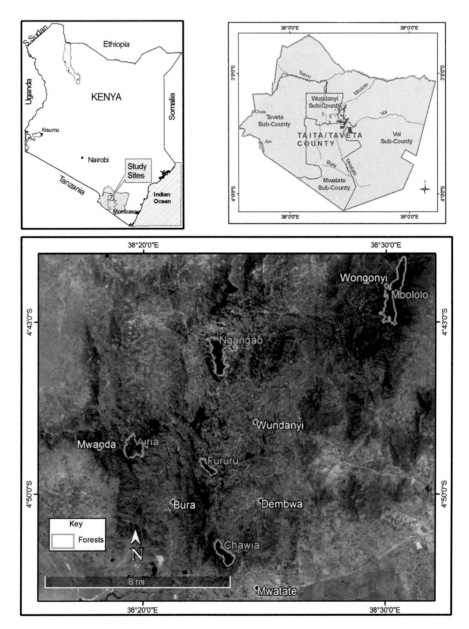

Fig. 2.1 Map showing the location of Taita Hills forests in Kenya (Source: KEFRI GIS Department)

Table 2.1 Basic information of the study area

Attribute	Description
Country	Kenya
Province	Coast
District	Taita Taveta
Municipality	Wundanyi
Size of geographical area (hectare)	1,708,390
Number of direct beneficiaries	9000
Number of indirect beneficiaries	137,618
Dominant ethnicity(ies), if appropriate	Dawida
Size of case study/project area (hectare)	769.8
Geographic coordinates (latitude, longitude)	3° 25' 0.12" S; 38° 19' 60.00" E

Table 2.2 Sampling design

Site	Number of villages	Population	Number of respondents	Number of respondents by socio-economic status			Respondents' representation (%)
				Poor	Middle income	Rich	
Chawia	4	12,098	35	18	11	7	14.1
Fururu	5	15,122	44	22	13	9	17.6
Mbololo	4	20,968	81	41	24	15	32.5
Ngangao	6	16,740	49	25	14	10	19.5
Vuria	6	13,990	41	21	12	8	16.3
Total	25	85,907	250	127	74	49	100

The number of respondents was proportional to the population size (Source: Field survey data, KEFRI research project)

2.2.2 Data Collection

Qualitative and quantitative surveys were used to establish the conservation programmes responsible for keeping the Taita Hills forests intact so as to ensure the provision of goods and services to the local communities. This approach provided breadth and depth of understanding and corroboration, while offsetting the weaknesses inherent in using each approach alone (Creswell and Plano Clark 2007). Participatory Rural Appraisals (PRAs) methods were used: key informant interviews were held in each community, and focus group discussions (FGDs) were conducted as part of both the qualitative and quantitative surveys.

Stratified random sampling was used to select respondents from different levels of socio-economic status based on income levels and general living standards which were determined by the communities during FGDs. The number of interviewees chosen was proportionate to the population of each of the sites (Table 2.2). Information on population in each of the villages was obtained from the 2019 Kenya Population and Housing Census report (KNBS 2019). Semi-structured questionnaires were used to collect data from 250 respondents in 25 villages surrounding the forest fragments. Twenty-five FGDs, one in each village, were also held with key informants (including village elders, members of local CBOs and NGOs, youth, women and experienced indigenous farmers) who are actively involved in conservation activities.

2.2.3 Age and Gender Profile of the Respondents

The majority of the respondents were middle-aged, accounting for 44%, followed by youth (34%), while the elderly was the least interviewed at 4% (Table 2.3). The youth and middle-aged adults are crucial stakeholders in present and future natural resources management in the Taita Hills landscape. Overall, 64% of the people interviewed during the survey were women.

2.2.4 Settlement History of the Respondents

Four different groups of respondents were identified based on the number of years lived on their present farms and/or origin and settlement history. Most of the respondents had either lived all their lives on the same farm or moved within the same sub-location (Table 2.4).

Table 2.3 Age composition of respondents (Source: Field survey data, KEFRI research project)

Age	Number	Response (%)
Less than 21	15	6.0
21–35	85	34.0
36–50	110	44.0
51–70	30	12.0
Greater than 70	10	4.0

Table 2.4 Settlement history of respondents in the study area

Years lived on present farm/origin	Number	Response (%)
Less than 20/outside Taita Hills	15	6.0
All their lives on the same farm	75	30.0
Less than 20/same sub-location	95	38.0
Less than 20/different sub-location	65	26.0

Source: Field survey data, KEFRI rescarch project

2.3 Results

2.3.1 Initiatives for Sustainable Conservation of Fragmented Forests of Taita Hills

The survey asked respondents about various initiatives being implemented by local communities to conserve the fragmented forests and associated biodiversity and support local livelihood systems. The most common responses were selected and listed (Table 2.5).

The most common conservation initiative found is the establishment of agroforestry belts on the forest edges to provide alternative sources of wood products and protect the forests from further encroachment, as reported by 34.8% of the respondents. The next most common was establishment of community tree nurseries for seedlings production (20.4%), followed by establishment of village conservation committees (16.8%), bee-keeping (15.6%), ecotourism (8.0%) and butterfly farming (4.4%).

2.3.1.1 Establishment of Agroforestry Belts

Ten-metre wide agroforestry belts have been established on private farmlands belonging to households adjacent to the forest fragments in order to ameliorate the adversarial micro-climatic conditions created on the forest edges by the edge effect. They provide an alternative source of wood products and protect the forest fragments from further encroachment. The amelioration of micro-climatic conditions on the forest edges has created favourable conditions for secondary forest growth, thus enhancing the resilience of the forest fragments to edge effect and contributing to the conservation of biodiversity, which could have otherwise been lost due to adversarial environmental conditions created by the edge effect. The belts are made up of mainly *Grevillea robusta*, an exotic tree species that was introduced in the area in the late 1970s and is a popular species for agroforestry because it grows very fast and, therefore, is intercropped with food crops such as maize and beans in the typical

Table 2.5 Initiatives developed by local communities to enhance conservation of fragmented forests of Taita Hills (Source: Field survey data, KEFRI research project)

Conservation initiative	Number of individual responses	Response (%)
Establishment of agroforestry belts	87	34.8
Establishment of community tree nurseries	51	20.4
Establishment of village committees to oversee conservation activities	42	16.8
Bee-keeping	39	15.6
Ecotourism	20	8.0
Butterfly farming	11	4.4

small farms owned by local communities. Moreover, the leaves of *G. robusta* provide abundant quantities of mulch, which accumulates to a depth of 30–40 cm and protects the soil in addition to enhancing soil fertility. Intercropping of food crops with *G. robusta* has improved crop yields due to enhanced soil fertility, thus ensuring food security. The belts have reduced illegal activities in the forests due to the availability of alternative sources of wood products such as firewood and timber, which prior to the belts' establishment were being extracted from the natural forests. *Albizia gummifera* is another tree species planted in agroforestry belts. Although *A. gummifera* is an indigenous species, some farmers prefer it for its ability to fix nitrogen in the soil and improve fertility, which is important in enhancing crop yields.

Related initiatives to conserve biodiversity in Taita Hills forests are also being undertaken by Nature Kenya, a local conservation NGO that has leased land from local farmers neighboring the Vuria forest and planted indigenous tree species that provide a favorable habitat for *Taita apalis* and *Taita thrush*, two bird species that are threatened with extinction, in an aim to save these bird species. Establishment of agroforestry belts has contributed immensely to provision of ecosystem goods and services that sustain local livelihoods, enhanced community adaptation to climate change, increased food production due to the net effects of agroforestry species on soil fertility, and enhanced biodiversity conservation, all of which have resulted in transformative change in the landscape as discussed later.

2.3.1.2 Establishment of Community Tree Nurseries

Community nurseries have been established in Chawia, Ngangao and Vuria to raise seedlings of native tree species for restoring degraded forest areas within the forest fragments (Fig. 2.2). The nurseries also raise seedlings of *G. robusta*, which is widely grown by local communities on their farmlands. The community nurseries produce about 50,000 seedlings annually, of which 10,000 are indigenous tree species while 40,000 are *G. robusta*. About 80% of the total tree seedlings produced are sold to stakeholders involved in conservation for planting in the forests and on farms, while 20% is planted by members of the community nurseries both in the forests (indigenous species) and on private lands (exotic species).

Indigenous species raised in community tree nurseries include *Prunus africana*, *Nuxia congesta*, *Ficus sycomorus* and *Albizia gummifera* because of their conservation and market value. These species are valuable for their timber and firewood, and the bark of *P. Africana*, which has medicinal value, can be sold for about 2.0 USD per kilogram. The species also have high conservation value as they protect water catchment areas. Other valuable indigenous tree species found in the community tree nurseries are *Ficus thonningii* for firewood and medicine, and *Newtonia buchananii*, *Syzygium guineense*, *Strombosia scheffleri* and *Maesopsis eminii* for timber. *Maesa lanceolata* and *Ocotea usambarensis* have potential market value as firewood or charcoal and therefore are also being raised in community nurseries for planting for fuelwood purposes. Seedlings of indigenous tree species are sold at 0.50 USD per

Fig. 2.2 Community tree nursery at Chawia forest (Source: Field survey data, KEFRI research project)

seedling, while exotic species are 0.10 USD per seedling. The community nurseries generate about 4000 USD from the sale of indigenous trees seedlings and 3200 USD from exotic tree species' seedlings per year. In total, the three community nurseries generate 7200 USD annually.

A proportion of the income (20% or 1440 USD) is ploughed back into conservation activities including increase of seedlings production and support for restoration of degraded sites within the forest fragments. The remaining amount of 5760 USD is shared among the 45 households with membership in the community nurseries. Consequently, each household receives 128 USD annually from the sale of seedlings raised in the community nurseries. Community nurseries contribute immensely to supplementing household income, conservation of natural forests through provision of seedlings for rehabilitation of degraded areas, provision of wood products including fuelwood thus reducing human pressure on natural forests, and preservation of indigenous knowledge associated with use of herbal medicine through propagation of herbal tree species.

2.3.1.3 Formation of Village Conservation Committees

The majority of the traditionally-protected small forest patches and sites that are important for conservation of water catchment areas are not gazetted. These small forest patches, traditionally known as *mitengo*, are located on private farms, trust land or on public land, for instance on riverfronts or roadsides. Thus, their management and conservation are mainly in the hands of local residents. Village conservation committees have been established to oversee conservation activities for the *mitengo*, and thus play an important role in forest management at the village level. The village conservation committees report incidences of destruction or encroachment into the small forest patches to administrative chiefs. Livestock grazing is not allowed in the *mitengo*; culprits are arrested and fined by local chiefs. The committees are also in charge of restoration of degraded areas within the *mitengo*. Thus, the committees spearhead the raising of seedlings in group and private nurseries and tree planting, in addition to raising community awareness on the importance of protecting and conserving these small forest patches, which are hubs for biodiversity conservation.

The committees also maintain and enhance social values traditionally attached to stable ecosystems, i.e. *mitengo*. The committees work with relevant authorities including the Kenya Forest Service (KFS)[2] and clan elders to ensure that the boundaries of the small-sized community forests are not altered. The traditional *mitengo* management system, which involves the village conservation committees' restrictions on activities that impact negatively on biodiversity, has safeguarded these fragile forests from illegal activities hence protecting many plant and animal species, most of which are endemic to these forests, and are endangered. Additionally, this traditional management system has ensured that the community forests are well conserved for provision of ecosystem services such as maintenance of water catchment areas and control of soil erosion, which is a challenge in the hilly terrain of the Taita Hills landscape. Employing traditional resource governance systems in the management of these *mitengo* has also resulted in preservation of cultural values key for sustainable management of the forests and their associated biodiversity.

2.3.1.4 Bee-Keeping

Bee-keeping is traditionally of considerable importance to the Taita community (Fig. 2.3). The main local use of honey is for the manufacture of traditional beer for important ceremonies. Honey is also used for medicinal purposes or taken with food. Bee-keeping as a modern enterprise was introduced in Taita Hills in the mid-1970s. Currently, there are 17 bee-keeping groups in the Taita Hills forests' landscape, with a total of 132 modern beehives installed in the five forest fragments.

---

[2]A government agency responsible for the conservation, sustainable development, management and utilisation of Kenya's forest resources for the equitable benefit of present and future generations.

Fig. 2.3 Bee-keeping in the forest (Source: Field survey data, KEFRI research project)

Each of the modern beehives produces about 15 kg of honey per harvest, worth 60 USD. This is an increase as compared to the 3 kg per harvest of the traditional log hives that were being used before the groups switched to modern beehives. The traditional log hives fetched only 12 USD per harvest, about 20% of what the modern beehives are generating for the local community members engaged in bee-keeping. The total income from honey per harvest is 7920 USD per harvesting season. There are 85 households involved in bee-keeping, and hence the income per household generated from honey per season is 93.20 USD. There are three harvesting seasons per year; accordingly, this translates to a household income of 279.60 USD annually. This household income from honey is substantial considering the fact that bee-keeping is not a full-time activity, and the majority of households largely depend on small-scale farming as their main economic activity.

The honey produced in Taita Hills is in high demand in the local, regional and national markets, and has substantially increased the income of local community groups, enhanced their food security and created employment for many local people along the value chain including the producers and vendors. The community members adjacent to the five forest fragments have therefore embraced bee-keeping as a non-consumptive and deforestation-free forest use to generate alternative income instead of encroaching into the forests to undertake illegal logging that is detrimental

Fig. 2.4 Tourist campsite at Ngangao forest (Source: Field survey data, KEFRI research project)

to biodiversity conservation. Furthermore, maintaining the high population of bees in the Taita Hills landscape promotes efficient pollination in the agricultural systems where the forests are embedded, leading to conservation of agrobiodiversity for traditional crops such as sorghum, millet, pigeon peas, green grams, and cowpeas, which also enhances food self-sufficiency.

2.3.1.5 Ecotourism

The value of biodiversity in Taita Hills forests is harnessed to support livelihoods and conserve the forests through ecotourism. Taita Hills have good ecotourism potential due to the unique forest flora and fauna diversity, scenic views, interesting culture of the Taita people and central location within the coastal tourism circuit. Ngangao, Vuria and Chawia forests are the leading destinations for ecotourism due to the rich diversity of flora and fauna including endangered bird species (*Taita apalis* and *Taita thrush*). These forests have campsites owned and run by a local community-based conservation group called Dawida Biodiversity Conservation Group (DABICO), where the tourists come and stay and pay the community group fees for camping and tour guiding (Fig. 2.4). DABICO has selected, recruited and trained tour guides from the local villages who take tourists around the forests and provide them with information on the flora and fauna diversity as well as the history and culture of Taita people.

Moreover, communities in Ngangao forest have constructed ecotourism *bandas* (traditional huts) and a restaurant through support from the Community Development Trust Fund (CDTF) (Fig. 2.4). The Ngangao forest has the highest ecotourism potential because it houses all the three endemic bird species and other avifauna that can easily be seen. The forest is also easily accessible. Plans to establish community-based ecotourism infrastructure in Mbololo forest are underway, where a community forest association (CFA) is in the process of launching an accommodation business in the form of homestays for tourists. The Mbololo forest hosts the endemic African violet (*Saintpaulia teitensis*) besides having unique scenic characteristics, and can offer great hiking experiences for nature enthusiasts.

Local communities consider ecotourism important as it generates income and offers the local youth job opportunities. A proportion of income from ecotourism is used to facilitate conservation and socio-economic development in Taita Hills. Tree nurseries have been set up using part of the income generated from ecotourism, and community groups are raising mainly indigenous tree species for replanting in degraded sites within the forest fragments. Income generated from ecotourism is also used to buy food for households during the dry season when there is insufficient food. Ecotourism has helped reduce non-sustainable uses of forests like cutting of trees for timber and charcoal production. Likewise, ecotourism has raised the value of the indigenous forests through improved conservation of biodiversity, and contributed to preservation of cultural heritage, one of the key attractions for tourists.

2.3.1.6 Butterfly Farming

The Taita Hills forests host diverse species of butterflies. A youth group in Chawia forest is engaged in butterfly farming (Fig. 2.5). The youth group, which consists of eight members, rears and sells butterfly pupae as one of its core livelihood activities. The group is also involved in the planting of indigenous tree species that are host to the butterflies. Conserving the forest by maintaining high flora diversity is important because the butterfly species found in the area are dependent on it. Likewise, the group considers conservation of Chawia forest a priority as it is directly linked to their livelihoods. The youth group has a tree nursery that raises about 5000 seedlings annually, mostly indigenous tree species preferred by butterflies that can host more than two butterfly species. The seedlings raised are planted in degraded sites within the forest as part of the group's forest restoration initiative. Additionally, each member of the youth group has to plant at least 50 trees annually on their farms to provide an alternative source of wood products.

The group raises butterflies starting with butterfly eggs, which hatch into caterpillars. The caterpillars are fed on the leaves of specific host tree species until they turn into pupae, which are sold to generate income for the group. More than 25 species of butterflies are found in Chawia forest, including *Cymothoe teita* and *Papilio desmondi teita*. These species are endemic to Taita Hills. Some species of

Fig. 2.5 Butterfly farming on the edge of Chawia forest (Source: Field survey data, KEFRI research project)

swallowtails and pansies are also found in Chawia forest. On average, the group sells about 200 butterfly pupae per month, translating to about 110 USD, or 1320 USD annually. Hence, each member receives 13.80 USD per month, an income that members of the community group use to supplement other income sources in order to meet their livelihood needs without engaging in destructive activities like illegal harvesting of trees for wood products such as timber and charcoal. The pupae are sold to the Kipepeo Butterfly Centre in Gede Kilifi County, which in turn exports to overseas markets, mainly in Europe, for live exhibits. Although the group rears two butterfly species endemic to Taita Hills forests (*Cymothoe teita* and *Papilio desmondi teita*), they do not sell their pupae. Rather they release the adults of the endemic butterfly species into the wild to increase their population and conserve the existing biodiversity of butterflies, which are important pollinators in the natural forests and agro-ecosystems in Taita Hills. Consequently, butterfly farming enables

the community group to earn income that sustains their livelihoods and also participate actively in the conservation of the Chawia forest, a win-win situation.

2.3.2 Indicators for Assessing Transformative Change

Five indicators have been developed by the local community to assess transformative change attributed to their conservation initiatives. In this particular case, transformative change refers to positive cultural change among local communities
towards conservation of forests and associated landscape through sustainable
utilisation practices that impact positively on local livelihoods and biodiversity
conservation. The indicators are: capacity building; replicability; governance and
social equity; livelihoods and well-being; and innovations (Table 2.6).

Increased capacity and capability of community groups to effectively participate
in conservation activities provide the foundation for upscaling the activities to cover
wider areas, and are used as a measure for long-term sustainability. Improved
livelihoods and well-being of the local people in terms of income, food security,
health and education are evidence for transformative change attributed to the robust
conservation programmes put in place by the local community. Greater participation
of women and youth in conservation activities enhances social equity and
strengthens the local governance system. The increased number of women and
youth involved in conservation and nature-based business ventures could be a
good indication of improved governance and social equity in the management of
the landscape. Local natural resources-based innovations incorporating traditional
resource governance systems play a key role in supporting concrete conservation

Table 2.6 Indicators for assessing transformative change in the Taita Hills landscape (Source:
Field survey data, KEFRI research project)

Indicators	Detailed description
Capacity building	The initiative has provided communities with skills and knowledge regarding conservation activities hence promoting the upscaling of best practices to the entire landscape
Replicability	The initiative has led to the increase in successful on-the-ground conservation and livelihood activities that are being replicated across the landscape
Governance and social equity	The initiative has ensured greater participation of women and youth in conservation activities and decision making on the way the landscape should be managed
Livelihoods and well-being	Income generated from the nature-based enterprises support the community in catering for health and education needs. Food security has also been enhanced due to improved crop yields attributed to better land management practices through agroforestry technologies
Innovations	Newly emerging locally-driven innovations have enabled the community to react to changes like climate change and discover new opportunities like revitalising traditional crop varieties

programmes. Newly emerging locally-led innovations that have high positive impact on the conservation status of the landscape and are likely to transform the landscape in terms of biodiversity conservation could also be evaluated to provide insights on the progress of transformative change of the people, forests and food security status. To upscale successful initiatives, replicability is paramount. The increase in successful on-the-ground activities that are being replicated across the landscape plays a pivotal role in bringing about transformative change, hence providing a good measure to assess the positive impacts attributable to these community-led conservation initiatives.

2.4   Discussion

The people-led conservation initiatives described herein are considered as a seed for transformative change with positive impacts on biodiversity and the livelihoods of local communities. However, there are several challenges for realising transformative change. These challenges include: the lack of conservation culture among the youth leading to high levels of deforestation; weak CFAs[3] with limited capacity to effectively participate in forest management; erosion of traditional knowledge systems that has led to destruction of shrines and other sacred sites within the landscape; poor attitudes towards conservation as some community members believe that conservation is a preserve of the government; and the lack of enabling policies, which encumbers communities' participation in conservation, e.g. limits on farmers who cannot harvest trees on their private farms without a permit from relevant government authorities discouraging communities to effectively engage in agroforestry. Consequently, to achieve transformative change in the management and conservation of the Taita Hills landscape, these challenges need to be addressed through institutional and policy changes that encourage active community participation in conservation initiatives within the landscape. Furthermore, youths should be given opportunities to take on leadership roles in CFAs to enable them participate in conservation matters, and to allow for traditional knowledge to be revitalised through inter-generational transmission of knowledge from the older generation to the youth. Moreover, the capacity of CFAs should be enhanced to equip community members with the requisite knowledge and skills to promote concrete conservation activities within the landscape.

There are existing opportunities that the local community are leveraging on to promote conservation and preserve biodiversity. The increasing market demand for deforestation-free nature-based products such as honey, herbal medicine and butterflies, whose sustainable production is dependent on the rich biodiversity of Taita Hills landscape, incentivises the local community to conserve the existing flora and fauna diversity. Moreover, there is increased capacity of the local community in tree

---

[3]Community Forest Associations.

husbandry; youth and women's groups have acquired knowledge and technical skills on propagation of trees and agricultural crops, and established nurseries to raise tree seedlings for planting in the farms and forests. Declining soil fertility due to soil erosion, which is responsible for low crop productivity, is also encouraging farmers to venture into agroforestry in order to improve soil fertility and enhance crop productivity. The local community is also revitalising traditional knowledge systems associated with agrobiodiversity conservation to conserve both forest and agriculture biodiversity. Moreover, the local community is aware and sensitive to the impacts of climate change, and therefore several community-based organisations (CBOs) are actively engaged in tree planting as mitigation measures to climate change. Farmers are also reverting to cultivation of traditional crop varieties that are resilient to the impacts of climate change to ensure the yields are adequate for the food security of the local community.

Nonetheless, there were contrasting views among some stakeholders/community members that traditional knowledge is backward and should not be applied in the present times in the management of landscapes; instead, science-based knowledge systems should take precedence. However, the majority of the stakeholders supported integration of traditional resource governance systems into conventional conservation approaches for complementarity and synergy. Existing partnerships and networks bringing together government agencies, NGOs, CBOs and other stakeholders involved in the conservation of the Taita Hills landscape enhances synergy and complementarity, resulting in high impact of the conservation programmes at multiple levels. Through the partnerships and networks, successful conservation initiatives have been replicated in two more areas through a 'multiplier effect', greatly contributing to biodiversity conservation and improved livelihoods of the people. A series of multi-stakeholder and policymaker workshops have been conducted to share implementation approaches, lessons learned and success stories of integrated conservation and livelihood initiatives in the landscape with a view to upscaling these initiatives in other landscapes in Kenya and the wider East Africa region.

2.5   Conclusion

The interdependence of ecosystem sustainability and community livelihoods is demonstrated in the Taita Hills landscape through the multipurpose sustainable use of the forests by local communities. The local communities have developed integrated conservation and livelihood approaches which they employ to conserve the fragmented forests for enhanced biodiversity conservation and livelihood improvement. These include: establishment of agroforestry belts adjacent to the forest fragments to ameliorate the adversarial micro-climatic conditions created on the forest edges and provide alternative sources of wood products to reduce pressure on the forests; establishment of community tree nurseries to restore degraded sites within the forest fragments and generate income; bee keeping, butterfly farming and ecotourism for livelihood improvement and enhanced conservation; and formation of village conservation committees to oversee conservation activities in the forest fragments. These approaches have been instrumental in conserving the fragmented forest patches while improving local livelihoods, and should therefore be promoted and upscaled for maximum benefits to biodiversity conservation and local economies, as well as for enhanced ecological and economic sustainability. The key messages and lessons learned from this particular study are elucidated below:

- Integrating nature-based microenterprises that support local economies in the management of the landscape and associated natural resources incentivises communities to participate in conservation activities.
- Enhancing the capacity of local communities through training on requisite knowledge and technical skills is key to sustaining conservation efforts in the long-term.
- Traditional knowledge should be integrated with scientific knowledge for complementarity and long-term effectiveness to deliver multiple societal benefits, including conservation, production, and livelihood benefits.
- Engaging policymakers to support community-led processes (bottom-up approach) in conservation initiatives promotes buy-in/ownership by the communities, hence ensuring sustainability of the activities.
- Multi-stakeholder engagement in conservation programmes that brings on board government agencies, CBOs, NGOs and local communities enhances synergy and complementarity resulting in high impact of conservation programmes at multiple levels.

Acknowledgements The authors would like to thank the Kenya Forestry Research Institute (KEFRI) through the Forest Biodiversity and Environment Management (FB&EM) thematic area for funding this study. We also thank the community researchers for undertaking the data collection. Finally, we thank the local Taita community for contributing valuable information that was used to develop this publication.

References

Adriaensen, F., Githiru, M., Mwangombe, J., & Lens, L. (2006). *Restoration and increase of connectivity among fragmented forest patches in the Taita Hills, South-east Kenya* (CEPF project report).

Burgess, N. D., Butynski, T. M., Cordeiro, N. J., Doggart, N. H., Fjeldsa, J., Howell, K. M., et al. (2007). The biological importance of the Eastern Arc Mountains of Tanzania and Kenya. *Biological Conservation, 134*, 209–231.

CIDP. 2013. *The First Taita Taveta County Integrated Development Plan 2013–2017.*

Creswell, J. W., & Plano Clark, V. L. (2007). *Designing and conducting mixed methods research.* Thousand Oaks, CA: Sage.

Hall, J., Burgess, N. D., Lovett, J. C., Mbilinyi, B., & Gereau, R. E. (2009). Conservation implications of deforestation across an elevational gradient in the Eastern Arc Mountains, Tanzania. *Biological Conservation, 142*, 2510–2521.

KNBS. (2019). *Kenya population and housing census* (Vol. I). Nairobi, Kenya: Kenya National Bureau of Statistics.

Myers, N., Mittermeier, R. A., Mittermeier, C. G., Da Fonseca, G. A. B., & Kent, J. (2000). Biodiversity hotspots for conservation priorities. *Nature, 403*, 853–858.

Newmark, W. D. (1998). Forest area, fragmentation, and loss in the eastern Arc mountains: Implications for the conservation of biological diversity. *Journal of East African Natural History, 87*, 29–36.

Omoro, L. M. A., Pellikka, P. K. E., & Rogers, P. (2010). Tree species diversity, richness, and similarity between exotic and indigenous forests in the cloud forests of Eastern Arc Mountains, Taita Hills, Kenya. *Journal of Forestry Research, 21*, 255–264.

Pellikka, P. K. E., Lötjönen, M., Siljander, M., & Lens, L. (2009). Airborne remote sensing of spatiotemporal change (1955-2004) in indigenous and exotic forest cover in the Taita Hills, Kenya. *International Journal of Applied Earth Observation and Geoinformation, 11*, 221–232.

Rogo, L., & Oguge, N. (2000). The Taita Hills forest remnants: A disappearing world heritage. *Ambio, 29*, 522–523.

Wekesa, C. (2018). *Effects of forest fragmentation on forest cover change, tree species diversity and carbon stock in Taita Hills, Kenya'*. PhD thesis, Egerton University, Njoro, Kenya.

Wekesa, C., Kirui, B., Maranga, E., & Muturi, G. M. (2019). Variations in forest structure, tree species diversity and above-ground biomass in edges to interior cores of fragmented forest patches of Taita Hills, Kenya. *Forest Ecology and Management, 440*, 48–60.

Wekesa, C., Leley, N., Maranga, E., Kirui, B., Muturi, G., Mbuvi, M., et al. (2016). Effects of forest disturbance on vegetation structure and above-ground carbon in three isolated forest patches of Taita Hills. *Open Journal of Forestry, 6*, 142–161.

Wekesa, C., Maranga, E., Kirui, B., Muturi, G. M., & Gathara, M. (2018). Interactions between native tree species and environmental variables along forest edge-interior gradient in fragmented forest patches of Taita Hills, Kenya. *Forest Ecology and Management, 409*, 789–798.

Chapter 3
Degraded Landscape Transformed into Foodland and Woodland by Village Agroforestry

Nathalie Viviane Raharilaza

Abstract This case study shares the results and lessons learned from agroforestry practices to restore a degraded and abandoned landscape, the production of seedlings of native and endemic tree species for forest restoration, and a trial of autochthones species transplantation at the village level in Madagascar. Awareness-raising and facilitation carried out by the NGO team on landscape changes and their effects on local people's lives, food and natural resources, were the initial drivers of this process. A farmer led the landscape restoration experimentation by taking part of his poor, degraded land that had been long abandoned, and giving the green light to use it as a 'farmer field school'. The community decided to keep the other side of the field untouched to enable comparison. Community members learned from each other by periodically sharing experiences. Community capacity-building on family accounting, production and harvest management helped community members to make decisions regarding the choice of crops and landscaping types suited to their needs. The community started to see results from the third year and increased the landscaped areas to boost future production. Some native trees like *Harina* (*Bridelia tuleasneana*), a highly preferred tree usually harvested from the rainforest for building materials, adapted very well to the village. The villagers learned to plant them rather than harvest them from the natural forest. The commitment, patience and courage of the community, and their immense pride in what they have achieved, created a cascading effect leading to sustainability.

Keywords Forest · Tradition · Landscape · Food security · Demography · Agroforestry

N. V. Raharilaza (✉)
Ny Tanintsika NGO, Pres Lot 089/3606-02 Ivory Avaratra, Fianarantsoa, Madagascar
e-mail: nat.nt.fnr@moov.mg

© The Author(s) 2021
M. Nishi et al. (eds.), *Fostering Transformative Change for Sustainability in the Context of Socio-Ecological Production Landscapes and Seascapes (SEPLS)*,
https://doi.org/10.1007/978-981-33-6761-6_3

3.1 Introduction

Madagascar, a country in southern Africa located in the Indian Ocean east of Mozambique, is the fifth largest island in the world, with 25.6 million people inhabiting a land mass of 587,000 km². More than 80% of 200,000 catalogued species (fauna and flora) are endemic. Madagascar remains a critical priority for international conservation efforts. Over the past 60 years, 44% of Madagascar's natural forests have disappeared, and deforestation rates have accelerated since 2005. Total biodiversity loss between 1950 and 2000 was estimated at 42%, of which 9.1% was estimated to be from forest loss alone (Allnutt et al. 2008).

In 2010, the country had 4,628,241 ha of primary forest with 16,450,024 ha of tree cover. However, this has decreased gradually to 4,060,522 ha of primary forest, with only 13,772,127 ha of tree cover in 2018 (Velo and Zafitsara 2020). With lack of funding to support agriculture development, timber harvesting and traditional slash-and-burn agriculture (locally known as "*tavy*") constitute the largest destructive forces of primary forest in Madagascar (Cormier-Salem et al. 2005). This phenomenal calamity has led to biodiversity loss with the extinction of thousands of endemic species. Creation of agricultural or pastoral lands at the expense of deforestation in Madagascar is an ongoing environmental issue that has had a profound effect on water resources, landscape design and features, habitat loss and soil impoverishment, among other negative impacts (Andriananja and Raharinirina 2004; Moreau 2013). Despite having considerable natural resources, Madagascar's poverty rate is amongst the highest in the world.

"Against poverty, for nature," is the slogan of Ny Tanintsika (NT) (www.nytanintsika.org), an NGO that encourages strategies and actions to improve lives, livelihoods and nature conservation in local communities bordering the rainforest. To contribute to the Satoyama Initiative's vision and global effort to realise societies in harmony with nature, we share the lessons learned by this case study which shows results from the ground related to landscape restoration through agroforestry and forest restoration in a project carried out and supported by NT since 2013.

In the context of transformative change as proposed by IPBES (2019), this project strongly influenced the following levers: incentives and capacity-building; pre-emptive action; and decision-making in the context of resilience and uncertainty. The leverage points shown to be effective in the context of this community were: visions of a good life (especially a vision out of poverty); values and action; inequalities; justice and inclusion in conservation (especially proactive actions by community members themselves); technology, innovation and investment; and education and knowledge generation and sharing.

3.2 Overview of the Project Location

The project was conducted in the village of Ambondro, located within the Ambohimahamasina rural municipality. Ambohimahamasina is 45 km southeast of Ambalavao, in the Fianarantsoa Province of Madagascar's central highlands. (Fig. 3.1 and Table 3.1).

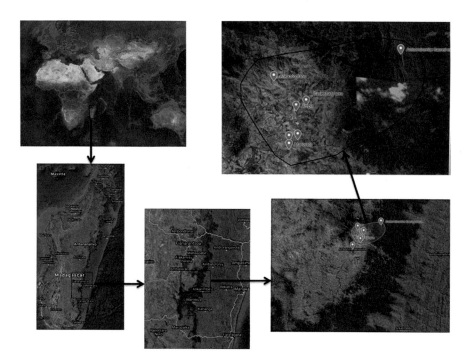

Fig. 3.1 Location and the surrounding landscape of Ambondro village

Table 3.1 Basic information of the study area

Country	Madagascar
Province	Fianarantsoa
District	Ambalavao
Municipality	Ambohimahamasina
Size of geographical area (hectare)	54,500
Number of indirect beneficiaries	37,418
Dominant ethnicity(ies), if appropriate	Betsileo
Size of case study/project area (hectare)	2330
Number of direct beneficiaries	500
Dominant ethnicity in the project area	Betsileo
Geographic coordinates (latitude, longitude)	21° 54′ 5.76″ S; 47° 12′ 56.88″ E

Fig. 3.2 Pierre RAMOMA, Guardian of the sacred mountain (Photo: H.F.RAHARILALAO)

In Malagasy, "Ambohimahamasina" literally means "at the scared mountain". The sacred mountain, named Ambondrombe, is believed to be where all spirits of the dead reside—the refuge of the souls of all Malagasy people's ancestors. Out of respect for the mountain of Ambondrombe on the part of the surrounding populations, the mountain's forest and biodiversity remains largely preserved. Pierre RAMOMA (Fig. 3.2), the village elder who holds the key to visiting the sacred mountain, is still alive and lives in Ambondro, our case study site. He is respected by the whole community, particularly young people. He tells them, "*I will not be here for much longer, so take responsibility. We need to protect our unique and common heritage*".

Bordering a rainforest stretching between Ranomafana and Andringitra National Parks, Ambohimahamasina has some stunning scenery, surrounded by numerous villages, rice fields and mountainous landscapes. Traditional livestock rearing is widespread, with the ownership of zebu cattle, a sign of wealth. The common practice is to allow herds to graze almost at will. To feed cattle, people burn land to let new grass shoots grow back (Messeri 2003). Basket weaving is traditionally practiced by women in this region and, most of the time, they harvest reeds and raw materials from the forest. People's lives depend on natural resources, and population growth has led to severe pressure on the landscape and natural resources.

These mountainous communities have found it difficult to determine a sustainable way to use the landscape and natural resources to meet their needs, which are growing day by day. At times, they have felt lost between external and internal views on use and conservation. Several environmental organisations work in the municipality with the common goals of forest protection, enhancing well-being and strengthening community resilience to climate change. However, the approaches and vocabulary used differ between organisations and local community—the local community cannot fully understand scientific explanations (e.g. carbon cycle, ecosystems). This is one of the reasons that prompted Ny Tanintsika to prioritise a more traditional approach based on traditional knowledge and inter-generational transfer of knowhow, and to choose Ambondro village as a study location. In Malagasy culture, leaving a legacy for future generations is considered a great responsibility, and we used this philosophy to sensitise people and encourage them to take responsibility to prevent deforestation and to leave forest and natural resources as a legacy for future generations.

Since the villagers were initially so dependent upon the rainforest to sustain their daily needs, the challenge was to produce food, raw materials, fruit trees and firewood close to the village. To do so, Ny Tanintsika supported the Ambondro community (Fig. 3.1) to develop agroforestry as a solution—not only for landscape restoration but also for food security (Randriatsara 2016; Randriamboavonjy 2015).

3.3   Challenges and Opportunities

The Ambohimahamasina rural municipality borders the rainforest corridor that runs along the eastern side of Madagascar. It is the remnant of a dense humid tropical forest that the Malagasy ancestors thought would never disappear; there was even a common saying *"when will the eastern forest run out?"* Nevertheless, this forest has been subject to excessive exploitation by people outside of the village over the last two decades, exposing it to unprecedented threats. Primary causes of forest loss include slash-and-burn culture and corruption (Dumont and Sinclair 2015).

Law 96-025, established on 30 September 1996, stipulated the transfer of natural resources management to the grassroots communities (Ratsifandrihamanana 2018). This law describes in depth the responsibility of forest administration and local community management. Basic community terms of reference detail the responsibilities to be carried out in the face of crimes and abusive exploitation of natural resources by internal or external persons, including immediate report to the forest administration that is in charge of prosecution and law enforcement. However, the weakness of the local management structure prevented it from playing an effective role as a Civil Society Organisation (CSO), and it easily fell into the trap of the big exploiters who are principal corrupters. Diverse NGOs and local government joined in an effort to enhance the capacity of community-based managers by providing some training on forest legislation, forest monitoring, communication and leadership skills. Figure 3.3 shows the inter-relationships between different stakeholders involved in this project. We note NT worked closely with these diverse entities to promote community development.

In order to conserve the rest of the forest, numerous nature protection organisations have supported communities to adopt environmentally-friendly behaviours towards sustainable natural resource management. Each organisation has developed its own strategy, but all of the actions have the common goal of forest conservation. Activities have included promotion of income generation initiatives, the establishment of community associations to manage local natural resources, the enforcement of rules in the case of illegal activity, and reforestation. Despite all these actions, poverty, combined with corruption, has contributed to a continual shrinking of the forest. Impacts are starting to be felt by local inhabitants and the landscape in general.

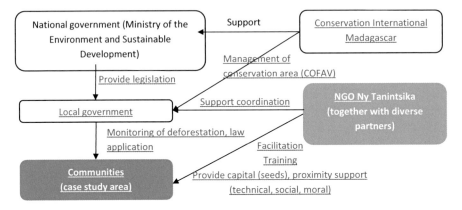

Fig. 3.3 Roles and relationships of stakeholders related to this project (Source: Author)

Ny Tanintsika, sister organisation of the Feedback Madagascar NGO (www. feedbackmadagascar.net), started working in the Ambohimahamasina municipality on projects to improve reproductive health and maternal and child health. Always concerned by the sustainability of actions and the well-being of the target community, Ny Tanintsika never hesitated to look for opportunities to develop partnerships and expanded the work in other development areas like water supply, women's empowerment, nutrition promotion, agroforestry and forest restoration. In light of our experience with these communities, different approaches have been adopted which take into consideration traditions, pride, management of life, family farm management, family planning, human rights, collective learning and intergenerational exchange. The following sections give more details on the approaches and processes that were followed as well as the results and lessons learnt.

3.4 Methodology

3.4.1 Selection of Community and Landscape to Target

Ny Tanintsika first defined certain criteria in order to select the site to be the target of the project on promoting agroforestry landscaping. In particular, a village with a high level of vulnerability and marginalism of community members, particularly vulnerability as regards to land, food, information, basic social services and limited access to other human needs, was sought.

The selected project site consists of mountain, hill and valley landscapes. As a result of deforestation and the practice of using fires to create grazing pastures, the landscape is highly degraded and has lost its main characteristics and functions in water retention (Messerli 2003). Erosion has also impoverished the soils. Because of this loss of productivity, the landscape was abandoned by the owners of the land. For the agroforestry restoration landscape, we chose a part of this land completely degraded and abandoned for a long time, in order to be able to show the changes.

3.4.2 Community Change Indicators

Indicators were chosen that are easy-to-use for the community members themselves, such as landscape change, improvements in agricultural yields, food source diversification, number of farmers adopting agroforestry techniques, number of fruit trees planted per household, the evolution of social cohesion and the emergence of community initiatives. We invited the community to establish a database for each indicator and also to measure and appreciate changes, thus facilitating community empowerment and the appropriation of change—a means towards sustainability.

Transformative change was assessed through the following indicators:

- Environmental indicators: Landscape design, landscape changes, and landscape features (soil occupation change, landscape management at the household and community levels, number of seedlings produced in tree nursery, number of trees planted, etc.);
- Economic indicators: Improved livelihoods, increased income, food diversification, and food security.
- Social indicators: enhancement and strengthening of social cohesion, community pride, and health services access.

3.4.3 Stakeholders' Roles

The role of each stakeholder was decided at the start of the process as follows:

- The Ny Tanintsika NGO ensured financial (e.g. pay for social outreach technician, travel expenses, and training costs), technical and social support to the community, and continual encouragement—acting as a mentor and advisor.
- The governmental agency was in charge of providing training on forest legislation and its application (www.environnement.mg).
- The private company BIONEXX (www.bionexx.com) provided *Cinchona sp* seedlings for experimentation as an agroforestry species. *Cinchona officinalis* is a medicinal plant used for the production of quinine, which is a fever-reducing agent. It is especially useful in the prevention and treatment of malaria. Other alkaloids that are extracted from this tree include cinchonine, cinchonidine and quinidine.
- The community members were the masters of their actions, their organisation, their choices and their decisions on the development of households and their village.

- Each household was a stakeholder, with its own household goal and action plan for a defined period.
- Women leaders were in charge of ten households each, providing training and monitoring.
- The agroforestry village technician implemented the model farm with the help of the community, taking note of all actions carried out in the field school.
- Two student groups from World Challenge UK (weareworldchallenge.com), Kingsmead College and Maidstone Grammar School, also provided help and some financial support, and worked with the community to build the granary storeroom during school holidays in 2018.

3.5   Activities and Community Initiatives

3.5.1   *Community Problem Analysis and Self-assessment*

A community meeting was held initially, with the participation of the entire community of Ambondro village—including elders, women, men, young people and even children. During the meeting, Pierre RAMOMA called also the *Ray amandreny lehibe* (the elder, the holder of the sacred mountain key, the owner of words, habits and customs) testified about the change in natural resources over time. He did not hesitate to say that a lot of change has happened over a short time—whether with regards to natural resources, food or lifestyle. He sounded the alarm to the whole community, especially young people, about the preservation of tradition, encouraging people to effectively take responsibility for stopping the scourge. He noted that society has changed and did not hide his concern for the loss of people's identity alongside the loss of natural resources. He invited people to imitate chameleons, who always look behind and forward simultaneously before taking a step, meaning that they must learn from lessons of the past, what was done, how people acted, what were the results of actions in daily life, whilst looking to the future—what will happen to our descendants, what can we do, what should we change?

At first, hearing about the changes in natural resources, food, the poor soil and low yield, the people were perplexed. They were convinced that conservation is important, but they did not know what to do and thought that nature conservation would require big resources that surely, they did not have. After a range of exchanges, the community made decisions at the end of this meeting that were to be translated into tangible actions. It was decided to carry out a collective experiment and to learn about the practice of agroforestry and landscape management using degraded soil and abandoned land as a farmer field school.

3.5.2   Natural Leader Emergence

A Malagasy proverb says, "An animal without a head cannot move" *(ny biby tsy misy loha ny tsy mandeha)*. After the community meeting, Pierre RASIJA (also called "Rapiera", Fig. 3.4) volunteered to pilot the experiment and collective learning for agroforestry practice and landscape management. He made available to the community part of his fields that were already degraded, sterile and populated by *Erica sp*, which he had abandoned for several years. Since then, he has played the role of Community Agroforestry Technician or TAC.

3.5.3   Learning by Practising and Believing by Seeing

Previous experience shows that it is not easy for farmers to understand a lot of theory taught in a classroom setting or to sit through long training sessions, and that learning through action is the best way of training. Capacity-building sessions in agroforestry management were held for community members in the village of Ambondro with practical training carried out directly in the field school (the degraded land). Villagers joined hands to develop the degraded land. Ploughing

Fig. 3.4   Pierre RASIJA, the natural leader in the degraded landscape (photo: Author)

and working the fields together, they had a regular schedule for gathering and for knowledge sharing. Participation in the field work was not compulsory; people were free to participate or not according to their will. This type of training helped everyone to deepen their understanding in different methodologies of landscaping, which they could then reproduce in their own fields. They learnt from their own experiences and accordingly, they were more likely to believe in the benefits of new techniques after seeing for themselves the differences in yields acquired. Training on topics such as leadership, yield management, health, family planning, water, sanitation and hygiene, and healthy food was also given during the process.

3.5.4 Women's Empowerment and Household Nutrition

In ancestral Malagasy culture, decision-making on matters of land use and management of natural resources is reserved for men, and women are merely required to adhere to the decisions made. Yet the kitchen is mostly reserved for women. We seised this opportunity as a gateway to involving women in decision-making in terms of improving family nutrition, a step towards empowering women.

As women are in charge of cooking and preparing food in the household, they are in the best position to apprehend the satisfaction of their families in terms of food, sufficient firewood and drinking water. Women were invited to carry out a simple analysis of the level of satisfaction with their household food situation, and encouraged to formulate their wishes for the future. As a consequence, two situations arose out of their analysis and reflections, which they translated into action plans for their respective families. Exchanges were held with women leaders on the daily management of food at the household level.

The results of this self-assessment show that households faced difficulties related to firewood scarcity (as there is a restriction on collecting from the natural forest), lack of food security, lack of money obligating them to borrow from rich people, shortage of seeds, and agricultural production insufficiency (amount produced insufficient to cover food year round).

The project's strategy was to enhance women's leadership, so it adopted a "cascade approach". This approach transfers competences. To boost women's leadership and empowerment, first four women leaders (Fig. 3.5) were identified by the community. The responsibilities of the women leaders were explained in depth to community members, and they were free to choose or design their women leaders according to their own criteria. The four women leaders identified by community members were trained in family life management, household and production management, leadership, healthy and balanced nutrition, reproductive health and community health. These women were empowered to be catalysers of change through this capacity building on life skills, and they were made responsible for transmitting the knowledge acquired to others, particularly for explaining the importance of the links between health, the well-being of the population, the countryside and the environment. After her training, each woman was responsible for monitoring and

Fig. 3.5 Razana, a woman leader of Ambondro village (photo: Author)

supporting another ten women. A self-assessment was carried out during the training on household food security to determine what is lacking and how problems could be resolved. Then we returned to the issue of land use, and participants learned from their experiences. Each household was invited to fix their goals for food security using the available land in their possession. Women leaders were trained to share with other women and help each other. A traditional proverb was used to promote community unification: "Those who are united form stones, and those who separate are sand". In general, participants were happy as this approach gave them something to be proud of, highlighting their importance for community development.

3.5.5 Sufficient Seeds for All

A lack of seeds, resulting from poor management of agricultural production, is one of the frequent problems encountered at the community level. At the beginning of the project, certified vegetable seeds were provided by the NGO to households who had decided to experiment with the landscaping agroforestry techniques

(Randriamboavonjy 2015). To facilitate the learning process, farmers were asked to compare results obtained from traditional and new agricultural techniques by dividing their land into two parts, thereby letting comparison of yield will show the difference.

A seed revolving strategy was adopted to resolve the problem of lack of seeds or unavailability of seeds for vulnerable households. Everyone who received free seeds from the NGO agreed to return three times the quantity at harvest time so that seeds could be given to three other farmers who requested help. In this way, all interested villagers had access to seeds and were able to boost production.

Seeing positive results from the pioneer participants, the number of households adopting agroforestry landscaping techniques increased progressively. To avoid seed consumption during lean periods, community members agreed to store their seeds in one place—a small space in RASIJA Pierre's house (owner of the farm school). However, as production increased, this space was no longer sufficient. Proud of the progress in their production, people requested help for the construction of a storage house (Fig. 3.6) for produce which was to double as a meeting and training room. Collaboration with groups of foreign students who finance community micro-projects during their holiday period resolved this need. The community now has a storehouse. The seeds (as well as some products) are stored until the arrival of the next cultivation period when they are given to all community members.

3.5.6 One Child, One Tree

"What you study is what you live". This was the slogan adopted to raise awareness amongst school pupils to apply and live their new knowhow in their households, daily lives and communities. NT had the prior experience of working with the Ministry of Education (www.education.gov.mg) to implement the "Child for Community" approach, which brought positive impacts, particularly in relation to promoting WASH (water, sanitation and hygiene) at the school level. For this approach, the children served as the messengers between school and their respective families, the village and other children. Thus, this same approach was used in relation to environmental protection, particularly tree-planting. Pupils were asked to each bring a sampling of a fruit tree that they found growing near their home (Fig. 3.7), to plant during a tree-planting operation held on National School Day on the school grounds (Fig. 3.8). Each pupil also brought one basket of compost. This helped to promote environmentally friendly behavioural habits adapted to the local context, such as preserving fruit seeds rather than throwing them away after eating fruit. The effort also sought to promote the mindset of finding local solutions rather than relying on purchasing of trees. There was close collaboration with the parents of students, who were in charge of digging the holes for tree-planting.

Fig. 3.6 Seed storage in the store room (photo: Author)

3.5.7   Natural Forest Restoration

Restoring degraded landscapes and developing landscapes around villages are challenges faced in meeting the daily needs of households and the community in terms of food and firewood. These basic needs contribute to forest loss. In addition to village agroforestry landscape planning activities, forest restoration is another major challenge (Dupin 2011). Pierre RASIJA, the farm school owner and also the nature conservation advocate in charge of the tree nursery, and Ndrasana, a young person from the community, were provided with training in phenological monitoring and seed collection from indigenous forest trees, the pre-treatment of these seeds (including seed selection, drying, soaking, etc.), their sowing and the production of indigenous seedlings in nurseries. The species that were chosen are those most used and liked by the community.

Fig. 3.7 Pupils from Ankarinomby Primary School, in Sahabe, with their own fruit-tree seedlings (photo: H.F.RAHARILALAO)

During the tree planting season, the communities organised themselves for the transportation and transplantation of native plant seedlings in the forest, sharing meals together during the work. This was a sign of their unity and commitment to forest restoration, a key part of their heritage.

3.5.8 Local Support and Partnership Development

The Ny Tanintsika community outreach technician provides support to the community on a continual basis (Fig. 3.9). Her role involves sharing information with communities and giving technical advice (Dumont and Sinclair 2015) whilst respecting social rules and traditions (including rules set with the community for collaboration, or social customs in the area).

Ny Tanintsika sought other partnerships in order to build further community support and, in particular, to boost income generation. A trial Cinchona (*Cinchona officinalis*) plantation was the fruit of collaboration between the communities and the company BIONEXX/QUIMPEXX. This shrub (Fig. 3.10), known for its anti-malarial properties, is more adapted and grown in degraded fields for agroforestry and for income generation from the sale of its bark. The results from initial trials on

Fig. 3.8 "I plant fruit trees, I'm happy, I will apply this to my own home." (photo: H.F.RAHARILALAO)

the model farm have been very promising, with good survival and growth rates, and these cinchona providing shade to crops grown below them. Other community members are now keen to plant these trees on their own land, and are in the process of negotiating for an expansion.

3.6 Results

3.6.1 Yield Improvement

After 4 years of the agroforestry landscaping experiment, communities saw a marked improvement in production with their harvest increasing threefold. This motivated them to continue efforts.

The staple Malagasy food is rice. Those who do not have enough ricefields are considered to be vulnerable people in the countryside. As previously mentionned, this is a vulnerable community that has very few ricefields. In the majority of cases, the most vulnerable work for others to get money to buy rice, and sometimes they

Fig. 3.9 Hortense, showing that we can develop and restore degraded land (photo: Author)

must borrow rice that they pay back in double during the harvesting period, causing their lives to remain difficult. Our intervention was designed to help them change their ideas about food, and to ensure food security by promoting the cultivation of other crops such as cassava. Community members grew cassava in their landscaping fields (wider area than rice fields), explaining the big change in yield. For rice farming, new techniques were applied after training in available fields in an aim to improve the rice yield, which produced some results (Fig. 3.11).

3.6.2 Diversification of Income Sources

Some community members have begun to diversify their sources of income for the betterment of their well-being in relation to natural environment. Marie Claire RASOANANDRASANA called Ra-Claire, single mother, 48 (Fig. 3.12), has reproduced the model field in her own fields and has been able to develop her turkey-farming activity. She is proud because she no longer has a debt problem. Previously she had to borrow money from loan sharks that she paid back at high interest rates after harvest. Now, her production is enough to feed her family, and a portion of income from sales has allowed her to buy young turkeys. She is

Fig. 3.10 Hortense, Ny Tanintsika Community Outreach Technician in the Cinchona plantation (photo: Author)

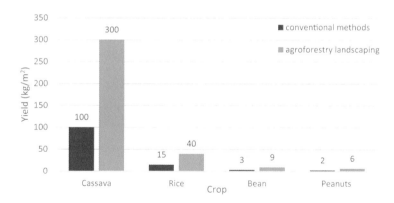

Fig. 3.11 Yield of major crops by conventional methods and landscaping methods Source: Data collection by Hortense (NT Community Outreach Technician)

multiplying the turkeys, and most importantly, she and her family built their own house in 2015.

Before this initiative, Raibida (a young man), one of the most vulnerable in the village, fed his family by stealing poultry and crops. Following the collective learning activities, he decided to work the land applying agroforestry landscaping

Fig. 3.12 Marie Claire RASOANANDRASANA fields, yields and turkeys (photo: H.F. RAHARILALAO)

techniques and stopped his bad habits. In 2016, he started to harvest his crops and money earned by selling corn and beans allowed him to buy a calf, and currently he can feed his family. He is an example of major change, and community members attest that "Raibida has changed".

The NGO team's role was advisory, always encouraging the whole community to have a clear vision for their future, define objectives for each activity or initiative, and continually improve and develop continually their livelihoods initiative. A special advice was given for the breeding activity such as taking care of the animal health and insuring their vaccinations on time. Finally, to help them to open their market vision we also provided an entrepreneurship development training.

3.6.3 Snowball Effect

Ny Tanintsika focused on promoting the efforts of joint work in Ambondro village, where the model farm school is located. A community sensitisation and information campaign on agroforestry was carried out. If a person was interested in learning more about agroforestry landscaping techniques, they were invited to come to Ambondro village and join in the efforts to practice. Women leaders that returned to their own villages ensured peer sensitisation, and they reproduced the model farm according to lessons learned there. Villagers, surprised by the success of the experiment, hastened to copy the techniques, even on a small scale, in their respective fields. That is exactly "**Believing by seeing**". Another factor that contributed to replication was people's enthusiasm for tree planting, in particular fruit tree planting and agroforestry techniques.

Currently, farmers in five different villages (Sahabe, Sahabe Est, Ampia, Ankazondrano and Marovato) have adopted these techniques, covering a total area of landscape of 2330 hectares. NT will continue to provide proof for the other villages, and will await the snowball effect, learning step by step how changes are adopted in communities (Fig. 3.13). Our previous experience has shown that changes adopted by the communities themselves are more sustainable.

3.6.4 Community Savings Initiative

Exchange with farmers from other localities was another positive factor that demonstrated the importance of farmer-to-farmer sharing of knowhow and experiences. A project promoting integrated action on health, population and environment was able to fund a visit of farmers from the Vohitsaoka municipality, another area in Ambalavao district, to Ambondro village. The visitors learned about agroforestry landscaping, and they also shared their experiences and success on a savings and loan group. This exchange prompted Ambondro villagers to create their own group savings, named VOAMAHASOA (literally *goodness seed*). Their goal is to help each other and to meet the repetitive financial difficulties encountered, another cause of their vulnerability. Currently, VOAMAHASOA, led by Rasoa RAJOMA (Fig. 3.14), has 20 members from three villages who meet every 2 weeks. During this meeting member each pay their own part, and they can also borrow money, paying back these small loans with a low interest rate (1%). To promote solidarity, loans related to medical consultations and the purchase of medicines are interest-free.

Fig. 3.13 Children in their field (Photo: H.F.RAHARILALAO)

3.6.5   Indigenous Trees in Village Fields

Rapiera (Pierre RASIJA), the village leader, the natural leader, the nature conservation advocate is also responsible for the production of native seedlings in the tree
nursery (Fig. 3.15). He works closely with Ndrasana, the indigenous forest seed
collector (Fig. 3.16). Whilst the majority of seedlings from the nursery are
transplanted into the natural forest for restoration, some of them are also planted in
village fields. Field experience has shown that *Harina* and *Rotrala* trees are most
adapted to the village. These trees are the most commonly gathered from the
rainforest for building materials.

3.6.6   Positive Thinking As the Key to Change

The main change in the community is the mentality. The people now have positive
thinking about life in general. They realised that they can accomplish something with
passion, unity and vision. Empathy, open-mindedness, confidence, observational
ability and energetic activity are the main characteristics of the natural leaders who
emerged.

Fig. 3.14 Rasoa RAJOMA, woman leader and treasurer of the VOAMAHASOA group (photo: Author)

3.7 Lessons Learned and Challenges

This study suggests the following lessons and challenges to be addressed:

- Sustainable changes come from the community. We do not impose them; we are their coaches.
- Consideration of community knowledge is a driver of change. Our task is to align them with modern techniques to achieve major results and impacts.
- The success of changes depends on the quality of facilitation and is related to facilitator behavior and capacity.
- Simple observation of changes is not enough. Scientific and statistical data are required to measure the success and evolution.
- The fight against corruption and deforestation remains a big challenge.
- Changing the subsistence economy into a market economy also remains a big challenge.

Fig. 3.15 Pierre RASIJA taking care of the tree nursery (photo: Author)

3.8 Conclusion

This landscape management project has demonstrated the multiple benefits of ecosystem restoration and the various actions taken to bring about transformative change. Not only has it improved ecosystem service provision, but it has also reinforced community solidarity. It has shown how human engagement is the key to change. Managing SEPLS at the micro-level can be used as an example of community climate change mitigation and adaptation, ecosystem restoration, and social equity and rights.

The satisfaction of households who followed instructions and reproduced the field school practices in their own fields represents a good start for sustainable change. When combined with other activities, important effects on livelihoods, ecosystem restoration and biodiversity conservation can be achieved.

"*Feed the earth, then it can feed humanity*" is the main message conveyed by the village leader. "*Prioritise humanity over all action*" is another. Empowering communities to make decisions concerning their own development is a key element of this project. It took at least 7 years to restore a degraded landscape.

Fig. 3.16 Ndrasana in his field (Photo: H.F.RAHARILALAO)

References

Allnutt, T. F., Ferrier, S., Manion, G., Powell, G. V. N., Ricketts, T. H., Fisher, B. L., et al. (2008). A method for quantifying biodiversity and its application to a 50 years record of deforestation across Madagascar. *Journal of the Society for Conservation Biology, 1*(4), 173–181.

Andriananja, H., & Raharinirina, V. (2004). Quels enjeux pour la durabilité et la gouvernance des ressources naturelles et forestières à Madagascar? *Mondes en développement, 127*(3), 75–89.

Cormier-Salem, M. C., Juhe-Beaulaton, D., & Boutrois J. (2005). *Patrimoine naturel au Sud*, IRD edition, Open edition book, viewed 30 September, 2020. https://www.editions.ird.fr/produit/9782709915601

Dumont, E. S., & Sinclair, F. (2015). *Guide technique d'agroforesterie pour la selection et la gestion des arbres au Nord-Kivu*. Nairobi, Kenya: World Agroforestry Centre.

Dupin, B. (2011). *L'agroécologie à Madagascar : analyse des conditions d'adoption paysanne de différentes techniques à partir de l'expérience d'AVSF*.

Feedback Madagascar. (2017). *Ny Tanintsika strategic plan 2018–2023*.

IPBES. (2019). Global assessment report on biodiversity and ecosystem services of the intergovernmental science-policy platform on biodiversity and ecosystem services. In S. Díaz, J. Settele, E. S. Brondízio, H. T. Ngo, M. Guèze, J. Agard, A. Arneth, P. Balvanera, K. A. Brauman, S. H. M. Butchart, K. M. A. Chan, L. A. Garibaldi, K. Ichii, J. Liu, S. M. Subramanian, G. F. Midgley, P. Miloslavich, Z. Molnár, D. Obura, A. Pfaff, S. Polasky, A. Purvis, J. Razzaque,

B. Reyers, R. R. Chowdhury, Y. J. Shin, I. J. Visseren-Hamakers, K. J. Willis, & C. N. Zayas (Eds.), *IPBES Secretariat*. Bonn: IPBES.

Law 96-025 1996, Relative a la gestion locale des ressources.

Messeri, P. (2003). *Alternatives a la culture sur brulis sur la falaise Est de Madagascar en vue d'une gestion plus durable de terre*. PhD de Geographie, Institut de Geographie, Universite de Berne Suisse. Viewed 30 September 2020, https://www.researchgate.net/publication/234091806_Alternatives_a_la_culture_sur_brulis_sur_la_Falaise_Est_de_Madagascar_strategies_en_vue_d'une_gestion_plus_durable_des_terres. January 2004.

Moreau, S. (2013). *Logique patrimoniale et conservation de la foret, exemple de la foret d'Ambondrombe, Sud-Betsileo Madagascar*. *Patrimoine naturel au Sud* (pp. 291–310). Marseille cedex: Edition IRD.

Randriamboavonjy, T. (2015). Inga, une plante utilisee pour la promotion de l'agroforesterie dans la NAP du massif d'Itremo, *KMCC blog-recherche et conservation de la biodiversite de Madagascar*, viewed 15 September 2020, https://teamkmcc.wordpress.com/2015/05/08/inga-une-plante-utilisee-pour-la-promotion-de-lagroforesterie-dans-la-nap-du-massif-ditremo/#:~:text=Itremo%20Massif%20NAP-,Inga%2C%20une%20plante%20utilis%C3%A9e%20pour%20la%20promotion%20de%20l'agroforesterie,NAP%20du%20Massif%20d'Itremo

Randriatsara, F. (2016). Trois ans de projet en agroforesterie mene par KMCC a Itremo', *KMCC blog-recherche et conservation de la biodiversite de Madagascar*, viewed July 7, 2016.

Ratsifandrihamanana, N. (2018). 'Gestion communautaire des ressources naturelles: l'avenir de Madagascar', World Wide Fund for Nature (WWF), viewed 10 janvier 2018, courrier des lecteurs, express de madagascar 15 decembre 2017.

Velo, N. M. A., & Zafitsara, J. (2020). The management of forest and deforestation, mitigation, and adaptation of climate change in Madagascar. *Open Access Library Journal, 7*(6), 1–19.

Chapter 4
Long-Term Tracking of Multiple Benefits of Participatory Forest Restoration in Marginal Cultural Landscapes in Himalaya

Krishna G. Saxena, Kottapalli S. Rao, and Rakesh K. Maikhuri

Abstract The literature is abound with references to the potential of indigenous and local knowledge (ILK) for sustainable landscape management, but empirical on-the-ground efforts that demonstrate this potential are still lacking. To identify interventions for improving the effectiveness and efficiency of forest restoration, participatory trials were set out in the Indian Himalaya, where per capita degraded land far exceeds per capita cropped/healthy forest land. Treatments were designed based on pooled indigenous and scientific knowledge taking into account farm-forest-livelihood interactions in cultural landscapes. The multipurpose tree-bamboo-medicinal herb mixed restoration plantation reached a state of economic benefit/cost ratio >1 in the eighth year and recovered 30–50% of flowering plant species and carbon stock in intact forest. The communities maintained but did not expand restoration in the absence of policies addressing their genuine needs and aspirations. Transformative change for sustainable restoration would include (1) nesting restoration in participatory, long-term, adaptive and integrated landscape development programmes, (2) formally involving communities in planning, monitoring, bioprospecting, and financial management, (3) assuring long-term funding but limited to the inputs unaffordable for local people, (4) stimulating the inquisitive minds of local people by enriching ILK and cultural heritage, (5) convincing policymakers to provide the scientific rationale behind policy stands, to support the regular interactions of communities with researchers, traders, and industrialists, to commit to genuine payment for ecosystem services in unambiguous terms at multiple spatial (household, village and village cluster) and temporal (short, medium and long-term) scales, and to

K. G. Saxena (✉)
School of Environmental Sciences, Jawaharlal Nehru University, New Delhi, India

K. S. Rao
Department of Botany, Delhi University, Delhi, India

R. K. Maikhuri
Department of Environmental Sciences, HNB Garhwal University (Central University), Srinagar, India

M. Nishi et al. (eds.), *Fostering Transformative Change for Sustainability in the Context of Socio-Ecological Production Landscapes and Seascapes (SEPLS)*,
https://doi.org/10.1007/978-981-33-6761-6_4

support long-term participatory action research for development of "landscape restoration models" in varied socio-ecological scenarios.

Keywords Bamboo · Biodiversity · Carbon stock · Medicinal plants · Farm · Integrated landscape management · Livelihood · Multipurpose trees · Payments for ecosystem services · Temperate

4.1 Introduction

Ecological restoration became a global priority at the turn of the twenty-first century, with the Convention on Biological Diversity aiming to restore at least 15% of degraded ecosystems by 2020, the Bonn Challenge to bring 350 million hectares of degraded and deforested land into restoration by 2030, and the United Nations declaring 2021–2030 as the Decade on Ecosystem Restoration. The envisaged targets can be achieved only through transformative change in current restoration approaches (IPBES 2019). A realisation of the key role of socio-economic and institutional factors in sustainable restoration (Chazdon et al. 2016; Reed et al. 2016) paved the way for community-centred forest landscape restoration and forest management (Baynes et al. 2015). Lately, the potential of indigenous and local knowledge (ILK) in overcoming technical barriers to restoration has also been recognised (Reyes-Garcia et al. 2019). Nonetheless, empirical efforts demonstrating this potential are rare (He et al. 2009; Stanturf et al. 2014). Forest restoration has been attempted largely through sponsored projects (Le et al. 2012; Burton 2014) and sustainability after cessation of funding has rarely been assessed (Rudel et al. 2005; Chazdon et al. 2016; Reed et al. 2016).

Himalaya is a biodiversity hotspot facing the entwined challenges of enhancement of livelihoods and environmental conservation. Forest restoration is a win-win option for conserving biodiversity, mitigating climate change and enhancing livelihoods (Brandt et al. 2017; Forest Survey of India 2019). We elaborate herein on the transformative change needed for sustainable forest restoration based on a participatory trial in a temperate socio-ecological production landscape.

4.2 Materials and Methods

4.2.1 Study Area

At the time of initiation of the study in 1991, Khaljhuni was a typical marginal, un-electrified village of the moist temperate region around the Nanda Devi Biosphere Reserve, a UNESCO World Heritage Site. The village territory comprised 68 hectares of terraced private farms and 436 hectares of community forests (Fig. 4.1 and Table 4.1).

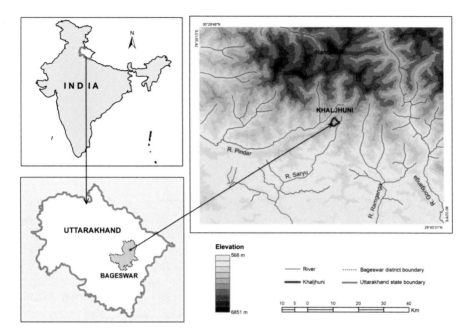

Fig. 4.1 Village of Khaljhuni in Central Himalayan Region of India

Table 4.1 Basic information of the study area

Country	India
Province	Uttarakhand
District	Bageswar
Municipality	n.a.
Size of geographical area (hectares)	5000
Number of indirect beneficiaries (persons)	10,000
Dominant ethnicity(ies), if appropriate	n.a.
Size of case study/project area (hectares)	600
Number of direct beneficiaries (persons)	340
Dominant ethnicity in the project area	n.a.
Geographic coordinates (latitude, longitude)	30° 2' 6.72" N; 79° 35' 52.80" E

Cultural heritage concerning the treatment of agricultural abandonment and purchase of foodgrains, as well as hiring of labour, from outside the village as evil omens, had checked agricultural expansion and strengthened social bonding at the time of the trial's initiation in 1991. Around 305 hectares of forest were intact and 131 hectares were severely degraded. The degraded forests were nearer to the dwellings than the intact forests. Degradation occurred because of lack of restoration after logging in 1910 by the Forest Department of the colonial government. The Department had created community forests to alleviate people's resentment against its disregard for their well-being and cultural heritage, which viewed logging as a bad omen, and had sanctioned income from wild non-timber forest products only to small farm holders, allowing only collection of edible/medicinal products from forests which checked storm flows and recharged springs. Confinement of ILK to passive forest restoration and prohibition of developing tree-crop mixed systems on forest land led the community to restore traditional practices in the intact forest and to use logged areas as free pasture, driving them to a severely degraded state.

4.2.2 Participatory Community Forest Restoration Trial

People's expectations and underpinning reasons for being against forest restoration were discerned from participatory discussions. People desired early and high income by growing medicinal herbs and understory bamboo. They dismissed the planting of trees due to long waiting periods for production of non-timber commodities, cultural restraints on income from timber, as well as hesitated to make cash or labour contributions to establish practices with uncertain returns. They seemed willing to reciprocate support for restoration addressing their expectations by (1) free sharing of ILK, deploying surplus physical resources and social capital, and maintaining/ expanding the trial once it reached a state of economic benefit/cost ratio of >1, and (2) accommodating the concerns of other stakeholders. They knew gregarious flowering induced mass mortality of bamboo but reconciled it with high productivity.

People gave up their initial reservations on planting trees after learning from participatory discussions that: (1) scientific knowledge lacked tested methods of cultivation of medicinal herbs/understory bamboo, (2) some multipurpose trees valued by them were likely to facilitate them, (3) organic branding of walnut/ honey will fetch premium prices, (4) payments for carbon sequestration were quite likely in near future, and (5) availability of tree products close to dwellings would save time/labour spent on collection from distant forests and would unleash opportunities for raising crop yields by eliminating tree-crop interferences. Ultimately, the community decided to establish a mixed plantation of trees, viz. *Aesculus indica* (horse chestnut), *Quercus leucotrichophora* (oak) and *Juglans regia* (walnut), a short bamboo (*Thamnocalamus spathiflorus*), and medicinal herbs viz. *Aconitum heterophyllum, Angelica glauca, Picrorhiza kurroa* and *Rheum australe* in an eight-hectare plot and maintain/expand it once economic outputs exceeded input costs. Financial support was terminated 7 years after planting when economic benefit/cost ratio crossed the mark of one (Rao et al. 1999), while participatory discussions and monitoring of the restored forest and competing land uses/economic activities continued.

4.2.3 Data Collection

Density of surviving plants, height/girth growth of trees/bamboo, biomass, soil carbon stocks and material inputs/outputs were measured after 1, 7, 10, and 20 years in the trial area and also in agricultural land, intact forests and degraded forests outside the trial area. All households were surveyed to assess changes in human/livestock population, land use-land cover, harvests and income. People's perceptions about the present trial and conservation development policies were ascertained from open-ended discussions, and researchers' deductions were informed in traditional open village assemblies in accordance with traditional norms (Rao et al. 1999; Maikhuri et al. 2000; Semwal et al. 2013).

4.3 Results

4.3.1 On-Site Outcomes and Impacts

4.3.1.1 Unanticipated Problems and Responses

Bamboo suffered gregarious flowering-induced mass mortality in the tenth year (2 years after cessation of funding) when it was at peak growth (80% of above-ground biomass). Economic loss from bamboo was far less than the gain from walnut, which coincidently started fruiting in the same year. At this time, people also felt a threefold increase in crop/beehive damage from bears and porcupines. People viewed this problem as a combined outcome of the decline in traditional collective mechanisms for dispelling harmful wildlife, facilitation of wildlife move-ment by bridges constructed over rivers/streams with road expansion, the wildlife sink function of the restored forest, and the focus on charismatic top carnivores (leopards/tigers) and their preferred ungulate preys in protected area management. Further, they envisioned that factors increasing pressure on intact forest (population growth, lack of cost-effective alternatives to timber/wood used for warmth, and cultural obligations of sharing resources with neighbouring communities that have lost their forests due to common amenities like electricity/road) would outweigh the ones decreasing pressure (access to subsidised cooking gas, modern medicine, vermicompost, subsidised food grains, and wage employment) in the near future. Comprehending their ILK, people transplanted wild saplings of *Alnus nepalensis* (Nepalese alder) in gaps for its fast growth in exposed nutrient poor soils, negative association with wildlife and timber value. This species was excluded in the initial planting when wildlife intrusion was limited, insect pests thriving on it threatened crops that met local food needs in the absence of external food supply, and there was yet a lack of envisioning of timber scarcity in future.

4.3.1.2 Species Selection, Performance and Ecological Recovery

People's selection of species was guided solely by direct economic benefits in the initial treatment and by both direct and indirect benefits in gap filling. Coincidently, species chosen for economic benefits were diverged in respect to leaf dynamics, nutrient requirements, shade tolerance and rooting patterns. ILK about fast growth of bamboo and alder in open habitats turned out to be true, though it altogether lacked underlying biological mechanisms. Naturally regenerated species shared just six percent of above-ground biomass after 20 years.

The rate of economic returns from the 20-year-old restored forest was higher than both the intact forest and cropland. On the other hand, the intact forest had better structure (canopy density, vertical stratification, and basal area) and higher regulating (carbon stock) and supporting (plant biodiversity) ecosystem services than the restored forest. Restoration over 20 years resulted in recovery of hardly 50% of flowering plant species and carbon stock in the intact forest. Species of high conservation value, the transplanted medicinal herbs, were altogether absent in cropland and untreated degraded forest (Table 4.2).

Time scale and accounting of biomass removals, below-ground organs and soil were significant determinants of carbon sequestration rate (Table 4.3).

4.3.2 Off-Site Impacts and Outcomes

There were two off-site impacts of the trial. First, three farmers started cultivating medicinal herbs on their private farms, which was boosted by financial support from National Medicinal Plants Board set up in 2000. The majority attributed non-adoption of this innovation to their exclusion from the scheme and hurdles in marketing. Secondly, 23 families leased out abandoned farms to a pharmaceutical company in 2005 for an annual income of Rs. 5000 ha^{-1} from the lease fee for 30 years. The company planted yew (*Taxus baccata*), which people used for preparing a health drink and for making the mainframes of houses. Extraction of the anti-carcinogenic drug paclitaxel (Taxol) by the company was outside the ILK domain. With the passage of time, people realised that the income from leasing was at the expense of land degradation (in the absence of manuring and weeding as yew was a stress tolerant species), their economic exploitation (rent being a fraction of the income to the company from Taxol), stagnation of ILK (exclusion of people from industrial processing), weakening of social bonding (the company negotiated with individuals) and illicit harvests from state forests at the company's behest. This realisation, a spin-off benefit from consistent monitoring, introspection, contemplation, social learning and adaptive responses learnt from the trial, led to a collective decision to defer new lease proposals. Further, conjecturing ILK as the foundation of Taxol discovery, people envisioned development of new commercial products from horse chestnut seeds and *Prinsepia utilis/Neolitsea pallens* seeds, used when crops

Table 4.2 Economic returns, vegetation structure, plant species richness and carbon stock in restored forests after 20 years, and other land uses (intact forest, degraded forest, cropland, and abandoned cropland) in the Khaljhuni socio-ecological production landscape

	Restored forest after 20 years	Intact forest	Degraded forest	Cropland	Abandoned cropland
Economic value of produce (Rs/ha/year)					
Timber[a]	–	700	–	–	–
Fuelwood[a]	747	1569	431	312	165
Fodder[a]	432	168	192	84	81
Bamboo—culms[a]	5550	178	–	–	–
Bamboo—seeds	–	200	–	–	–
Medicinal herbs	902	267	163	176	121
Lichens	–	97	–	–	–
Leaf litter[a]	–	137	–	–	–
Foodgrains	–	–	–	21,320	–
Fruits	27,360	240	0	360	0
All products	34,991	3556	786	22,252	367
Vegetation structure and ecosystem functions					
Canopy density (%)	40–60	>70	<5	<5	10–20
Vertical stratification	Upper canopy of 3–7 m tall planted trees, lower canopy of 1–2 m tall naturally regenerated trees and <1 m tall ground vegetation	Upper canopy of 15–25 m tall trees, lower canopy of 5–10 m tall trees and 1–2 m tall ground vegetation	Isolated 1–3 m tall trees and lack of stratification of tree canopy	Isolated 7–12 m tall trees and lack of stratification of tree canopy	Isolated 3–5 m tall trees and lack of stratification of tree canopy
Number of plant species					
Trees	8	21	2	5	6
Shrubs/bushes	3	12	5	3	3
Bamboos	1	3	0	0	0
Herbs other than grasses	7	27	14	5	11
Grasses	19	7	19	11	16
All species	38	70	40	24	36
Carbon stock (Mg carbon/ha)					
Below-ground	33.6	189.6	19	15.8	10.4

(continued)

Table 4.2 (continued)

	Restored forest after 20 years	Intact forest	Degraded forest	Cropland	Abandoned cropland
Above-ground	85	113.8	28.8	66.7	27.7
Total	118.6	303.4	47.8	82.5	38.1

Note: Provision of payments for regulating/supporting ecosystem services did not exist
[a]Social norms permitted sale of products only by small farm holders/weaker families to large farm holders/affluent families within the village

Table 4.3 Changes in carbon sequestration rates with progression of participatory forest restoration trial in village Khaljhuni, Central Himalaya, India

Carbon sequestration rate (Mg carbon ha^{-1} year^{-1})	Initial 7 years of restoration	8th to 20th year of restoration	Average over 20 years of restoration
Above-ground			
Excluding biomass utilised	1.06	0.75	0.86
Including biomass utilised	2.29	3.15	2.85
Below-ground[a]	3.55	2.39	2.80

[a]Includes carbon in complete soil profile

failed as a staple food and an edible oil source, respectively, and from *Juglans regia*, used in healthcare/dyeing/pest control.

4.3.3 Khaljhuni Cultural Landscape Over the 1991–2011 Period

Abandonment of cropping initiated in 2000 covered 29% of farm area by 2011 due to: (1) failure of ILK-based alder planting in checking wildlife menace and inability of government agencies in developing solutions to the problem, (2) lack of policy for compensating crop/honey losses due to wildlife, (3) land holdings becoming too small (0.2 ha per capita in 2011) to secure livelihoods and (4) decline in cultural values prohibiting agricultural abandonment and fostering collective wildlife control due to access to subsidised food grains and acculturation to improved accessibility (distance to road reduced to 6 km in 2011, from 14 km in 1991). Forest recovery failed in abandoned farms as a result of the open-access status of unleased farms and stunted yew growth in leased/protected farms.

Government agencies planted conifers on seven hectares of degraded community forest between 2006–2007, engaging local people as wage labourers. The plantation suffered a 100% mortality within a year due to fires set by the people themselves, who value herbaceous forage more than conifers. Administrative machinery was too weak to check illicit forest fires and private farm leasing.

People listed many reasons behind the failed expansion of the trial: (1) it could not mitigate the new wildlife menace problems, decline in the demand for bamboo handicrafts and competition/uncertainties in walnut/medicinal herb markets; (2) it did not satisfy new aspirations viz., freedom to harvest and market all forest products, establishment of value-added facilities within the village and increase in quantum of government support for restoration; (3) technological stagnation created a mindset of looking down upon the traditional labour-intensive production system; and (4) it could not compete with new opportunities for income from caterpillar fungus (*Cordyceps sinensis*) and government-funded infrastructure development projects in the absence of protected markets for restoration products and commitments to payments for ecosystem services. Agricultural abandonment nullified restoration-mediated carbon sequestration.

4.4 Discussion

4.4.1 Socio-Ecological Diversity

Subjective carving of community forests resulted in enormous variation in their extents, accessibility and ecological status. It is common to observe communities conserve intact forests and restore selectively logged forests vast enough to secure their essential needs by deploying their ILK and social capital centred around passive restoration. Because of cultural restraints on the timber trade, the intact community forests of Khaljhuni were more efficient in storing carbon (Buffum et al. 2008; Sharma et al. 2010; Pandey et al. 2014) and harbouring species of high conservation (Maikhuri et al. 2000; Adnan and Hölscher 2011) and ecological values (Bhadauria et al. 2012) than the ones managed by the communities free from such restraints. On the other hand, intact forest patches lacking religious value and too small to meet subsistence needs, recharge potable water sources, mitigate floods and to stabilise slopes, suffered from illicit logging by outsiders and then encroachment by local people (Semwal et al. 2004; Wakeel et al. 2005). Due to increasing integration into the mainstream, Khaljhuni people aspired for a quantum leap rather than an incremental rise in income, feasible only by active restoration lacking in ILK. Divergent from people's expectations, conventional restoration targeted forest area expansion to cover two-thirds the area of mountains and one-third the area of the country at the time of trial initiation in the early 1990s.

4.4.2 Environment-Knowledge-Culture-Policy Interlinkages

The initial demand for growing marketable herbs/short bamboo was a combined reflection of the people's aspirations for early and high income and their ignorance of scientific knowledge (Guariguata et al. 2010; Ashton et al. 2014) and the expectations of other stakeholders. Participatory discussions enriched ILK and motivated people to minimise trade-offs between immediate economic and long-term ecological benefits from restoration (Yami et al. 2013; Andrews and Borgerhoff Mulder 2018). Attention to non-material benefits and people's priorities, which were excluded in top-down restoration, enabled people to (1) conceive new ideas viz., alder planting in gaps in the restored plot and medicinal herb cultivation in private farms, and visualisation of ILK-based new commercial products, (2) revise/rectify based on new learnings reflected from inclusion of tree planting, deferring of new lease proposals and abandonment of cultural values forbidding income even from over-mature/dead/diseased trees, and (3) offer in-kind contributions which reduced treatment establishment (Rao et al. 1999) and monitoring costs (Evans et al. 2018). ILK has been increasingly recognised as an under-exploited resource and community participation as a means of cost-effective forest management and inclusive development worldwide after 1980.Yet, forest restoration planning, monitoring and financial management remains the state monopoly. Top-down restoration projects continue to thrive as a result of projection of their success based on achievement of planting targets rather than the magnitude of ecological recovery (Le et al. 2012; Dudley et al. 2018). These projects serve the vested interests of a few officials and elected representatives more than community well-being (Barr and Sayer 2012; Baynes et al. 2015). Likewise, bioprospecting remains the monopoly of mega-industries in the absence of strict enforcement of free, prior and informed consent procedures and ambiguous benefit-sharing provisions (Maikhuri et al. 2004).

Unlike conventional projects treating people solely as beneficiaries, this project involved them as responsible stakeholders owning all decisions, collecting scientific data and maintaining expenditure accounts jointly with the researchers. Following the traditional norms, final decisions were made by consensus in an open assembly steered by the community leaders and facilitated by the researchers (Rao et al. 1999). Whole-hearted community participation thus derived from researchers appreciating and enriching ILK and cultural heritage, recognising people's needs/aspirations, living with them, establishing a transparent expenditure accounting system and earnestly following the principles of co-management (Berkes 2009). Voluntary maintenance from the eighth year onwards testifies that people tend to honour their informal commitments and take collective action when they encounter anticipated shocks like bamboo flowering. Nonetheless, the trial succeeded in securing investment in restoration, but not in its voluntary expansion or upgradation. A quantum rise in effectiveness and efficiency of current forest restoration approaches is the key transformative change needed to arrest/reverse global deforestation and forest degradation. The present study unveils that this change is feasible from a set of concomitant policy changes for (1) harnessing and enriching ILK and cultural heritage, (2) motivating people to contribute surplus physical resources, (3) making people a formal party in designing and monitoring treatments and financial management, (4) rescuing people from unanticipated problems like excessive wildlife intrusions and economic exploitation by industries and traders whose mitigation is beyond their knowledge and capacity, and (5) nesting restoration in long-term and adaptive landscape management-cum-livelihood enhancement plans.

Policies have changed, but are far from the expectations of people and initiatives in other countries (Berkes 2009; Liu et al. 2016). Lack of holistic and integrated landscape management approaches advocated since the 1980s manifested in decline in competitiveness of forest restoration. With increasing education and awareness, people are becoming disgusted with policy elements lacking scientific and/or socio-economic rationale: (1) transfer of only 18% of forest area taken over from them in the nineteenth century and community empowerment just to regulate subsistence uses in an era of economic globalisation, (2) meager financial support for ILK-based treatments (Derak et al. 2018) and persistence of cultural heritage favouring equity, community solidarity and environmental sustainability (Maikhuri et al. 2000), (3) elongation of restoration project duration without raising the quantum of financial support, (4) arbitrary selection of households for availing government incentives, and (5) lack of commitments on community share in UN-REDD+ and other mechanisms of payments for ecosystem services. ILK on medicinal plants and new demand for them in national and overseas markets prompted policy uptake, but budget support was too low to induce any transformative change. Further, policy support was largely confined to cultivation in private farms, anticipating that wild populations would be restored. This was unsuccessful due to loopholes in the existing mechanisms of checking illicit harvests and their marketing (Rao et al. 2015). Failure of policy uptake for long-term support for participatory-adaptive restoration and for synergising forest conservation, forest restoration, agricultural development and socio-economic upliftment underscores a need for effective interactions among stakeholders: people, researchers, industrialists, traders and government officials. This failure seems to be the root cause of the lack of competitiveness of current forest restoration pathways, nullification of restoration-mediated carbon sequestration by agricultural abandonment-mediated emissions, and people's perceptions about the unfair sharing of benefits from new products.

Voluntary maintenance of the trial area despite poor survival of medicinal herbs and the ineffectiveness of alder in checking wildlife intrusions suggest that people value whole-hearted efforts more than their outcomes (Andrews and Borgerhoff Mulder 2018). Voluntary maintenance of the trial from the eighth year onwards was a sort of reciprocation for external support to the satisfaction of the people. Participation merely as wage earners resulted in unproductive investment in top-down restoration planting in the absence of a sense of ownership, responsibility and accountability among people and law enforcement (Schultz et al. 2012; Lyver et al. 2019; Reyes-Garcia et al. 2019). Discontentment among people due to neglect of their genuine concerns may turn into mass movements, like the ones in the 1920s forcing government to create community forests, in 1970s to ban logging even by its agencies in natural forests and to restore ecotourism, for example, in the Nanda Devi Biosphere Reserve (Maikhuri et al. 2000; Rao et al. 2015).

Assuming persistence of a net above-ground accumulation rate of 0.75 Mg C ha^{-1} year^{-1} observed at the end of the 20th year, it will take 210 years to restore C stock in the temperate intact Himalayan forest compared to 50–100 years in tropical forests (Marin-Spiotta et al. 2007; Wheeler et al. 2016; Lewis et al. 2019). Recovery in species richness was faster than carbon stock due to planting of a large number of species and minimal weeding. Active restoration is more expensive than passive restoration but a necessity in a biodiversity hotspot like Himalaya with intense limitations of propagule dispersal, viable soil seed banks and safe sites in shallow-gravelly soils on steep slopes. This study also underscores a need for treating cultural landscapes (Takeuchi et al. 2014; Reed et al. 2016) rather than just forests (Vanderhaegen et al. 2015) as spatial units, and for clarity on accounting of below-ground and harvested biomass carbon pools, and for time scale in determining payments for ecosystem services.

4.5 Conclusions

The present long-term location-specific study shows that nature and outcomes of participation of local people in forest restoration would vary in space and time depending on the socio-ecological conditions. Our field visits during the 2011–2019 period after systematically analysing the landscape for 20 years suggest the persistence of agricultural abandonment, restoration failure beyond the trial area, conservation of intact forest and the mindset of higher social status of urban livelihoods. The present study unveils the following practices for transformative change to achieve national or global targets:

- Nesting restoration in participatory long-term adaptive and integrated landscape and livelihood enhancement programmes;
- Coupling commodity production with bioprospecting, manufacturing and marketing;
- Formally involving communities in planning, monitoring and financial management;
- Assuring funding until restoration becomes an additional source of material/ income benefits to people;
- Limiting funding to inputs unaffordable by people with a condition of people-researcher joint monitoring after cessation of funding;
- Stimulating the inquisitive minds of people by informing them of the scientific implications of their ILK, as well as that evolved by other communities;
- Convincing policymakers to provide scientific foundations for stands/actions in open domain, and necessary support for the development of a network of "model restoration landscapes" co-managed by local people, researchers, traders and industrialists, and to commit a fair share to communities in payments for ecosystem services;
- Promoting condition/performance-based incentives/subsidies/rewards/compensation at multiple spatial (household, village, and village cluster) and temporal (short, medium, and long-term) scales.

References

Adnan, M., & Hölscher, D. (2011). Medicinal plants in old-growth, degraded and regrowth forests of NW Pakistan. *Forest Ecology and Management, 261*, 2105–2114.

Andrews, J., & Borgerhoff Mulder, M. (2018). Cultural group selection and design of REDD+: insights from Pemba. *Sustainability Science, 13*, 93–107.

Ashton, M. S., Gunatilleke, I. A. U. N., Gunatilleke, C. V. S., Tennakoon, K. U., & Ashton, P. S. (2014). Use and cultivation of plants that yield products other than timber from South Asian tropical forests, and their potential in forest restoration. *Forest Ecology and Management, 329*, 360–374.

Barr, C. M., & Sayer, J. A. (2012). The political economy of reforestation and forest restoration in Asia-Pacific: Critical issues for REDD+. *Biological Conservation, 154*, 9–19.

Baynes, J., Herbohn, J., Smith, C., Fisher, R., & Bray, D. (2015). Key factors which influence the success of community forestry in developing countries. *Global Environmental Change, 35*, 226–238.

Berkes, F. (2009). Evolution of co-management: Role of knowledge generation, bridging organizations and social learning. *Journal of Environmental Management, 90*, 1692–1702.

Bhadauria, T., Kumar, P., Kumar, R., Maikhuri, R. K., Rao, K. S., & Saxena, K. G. (2012). Earthworm populations in a traditional village landscape in Central Himalaya, India. *Applied Soil Ecology, 53*, 83–93.

Brandt, J. S., Allendorf, T., Radeloff, V., & Brooks, J. (2017). Effects of national forest-management regimes on unprotected forests of the Himalaya. *Conservation Biology, 31*, 1271–1282.

Buffum, B., Gratzer, G., & Tenzin, Y. (2008). The sustainability of selection cutting in a late successional broadleaved forest in Bhutan. *Forest Ecology and Management, 256*, 2084–2091.

Burton, P. J. (2014). Consideration for monitoring and evaluating forest restoration. *Journal of Sustainable Forestry, 33*, S149–S160.

Chazdon, R. L., Brancalion, P. H. S., Laestadius, L., Bennett-Curry, A., Buckingham, K., Kumar, C., Moll-Rocek, J., Vieira, I. C. G., & Wilson, S. J. (2016). When is a forest a forest? Forest concepts and definitions in the era of forest and landscape restoration. *Ambio, 45*, 538–550.

Derak, M., Cortina, J., Taiqui, L., & Aledo, A. (2018). A proposed framework for participatory forest restoration in semiarid areas of North Africa. *Restoration Ecology, 26*, S18–S25.

Dudley, N., Bhagwat, S. A., Harris, J., Maginnis, S., Moreno, J. G., Mueller, G. M., Oldfield, S., & Walters, G. (2018). Measuring progress in status of land under forest landscape restoration using abiotic and biotic indicators. *Restoration Ecology, 26*, 5–12.

Evans, K., Guariguata, M. R., & Brancalion, P. H. S. (2018). Participatory monitoring to connect local and global priorities for forest restoration. *Conservation Biology, 32*, 525–534.

Forest Survey of India. (2019). India State of Forest Report 2019, Ministry of Environment, Forests and Climate Change, Government of India, Dehradun, India.

Guariguata, M. R., Garcia-Fernandez, C., Sheil, D., Nasi, R., Herrero-Jauregui, C., Cronkelton, P., & Ingram, V. (2010). Compatability of timber and non-timber forest produce management in natural tropical forests: Perspectives, challenges and opportunities. *Forest Ecology and Management, 259*, 237–245.

He, J., Zhou, Z., Weyerhaeuser, H., & Xu, J. (2009). Participatory technology development for incorporating non-timber forest products into forest restoration in Yunnan, Southwest China. *Forest Ecology and Management, 257*, 2010–2016.

IPBES. (2019). *Summary for policymakers of the global assessment report on biodiversity and ecosystem services of the Intergovernmental Science-Policy Platform on Biodiversity and Ecosystem Services*. Bonn, Germany: IPBES Secretariat.

Le, H. D., Smith, C., Herbohn, J., & Harrison, S. (2012). More than just trees: Assessing reforestation success in tropical developing countries. *Journal of Rural Development, 28*, 5–19.

Lewis, T., Verstraten, L., Hogg, B., Wehr, B. J., Swift, S., Tindale, N., Menzies, N., Dalal, R. C., Bryant, P., Francis, B., & Smith, T. E. (2019). 'Reforestation of agricultural land in the tropics:

The relative contribution of soil, living biomass and debris pools to carbon sequestration. *Science of the Total Environment, 649*, 1502–1513.

Liu, S., Dong, Y., Cheng, F., Coxixo, A., & Hou, X. (2016). Practices and opportunities of ecosystem service studies for ecological restoration in China. *Sustainability Science, 11*, 935–944.

Lyver, P. O. B., Richardson, S. J., Gormley, A. M., Timoti, P., Jones, C. J., & Tahi, B. L. (2019). Complementarity of indigenous and western scientific approaches for monitoring forest state. *Ecological Applications, 28*, 1909–1923.

Maikhuri, R. K., Nautiyal, S., Rao, K. S., Chandrasekhar, K., Gavali, R., & Saxena, K. G. (2000). Analysis and resolution of protected area-people conflicts in Nanda Devi Biosphere Reserve, India. *Environmental Conservation, 27*, 43–53.

Maikhuri, R. K., Rao, K. S., & Saxena, K. G. (2004). Bioprospecting of wild edibles for rural development in the Central Himalayan Mountains of India. *Mountain Research and Development, 24*, 110–113.

Marin-Spiotta, E., Ostertag, R., & Silver, W. L. (2007). Long term patterns of tropical reforestation: Plant community composition and aboveground biomass accumulation. *Ecological Applications, 17*, 828–839.

Pandey, S. S., Maraseni, T. N., & Cockfield, G. (2014). Carbon dynamics in different vegetation dominated community forests under REDD+. *Forest Ecology and Management, 327*, 40–47.

Rao, K. S., Maikhuri, R. K., & Saxena, K. G. (1999). Participatory approach to rehabilitation of degraded forestlands: A case study in high altitude village of Indian Himalaya. *International Tree Crops Journal, 10*, 1–17.

Rao, K. S., Saxena, K. G., & Tiwari, B. K. (2015). *Biodiversity, climate change and livelihood in the Indian Himalaya: An overview*. Dehradun, India: Bishen Singh Mahendra Pal Singh.

Reed, J., Van Vianen, J., Deakin, E. L., Barlow, J., & Sunderland, T. (2016). Integrated landscape approaches to managing social and environmental issues in the tropics: Learning from the past to guide the future. *Global Change Biology, 22*, 2540–2554.

Reyes-Garcia, V., Fernandez-Llamazares, A., McElwee, P., Molnar, Z., Öllerer, K., Wilson, S. J., & Brondizo, E. S. (2019). The contributions of indigenous peoples and local communities to ecological restoration. *Restoration Ecology, 27*, 3–8.

Rudel, T. K., Coomes, O. T., Moran, E., Achard, F., Angelsen, A., Xu, J., & Lambin, E. (2005). Forest transitions: Towards a global understanding of land use change. *Global Environmental Change, 15*, 23–31.

Schultz, C. A., Jedd, T., & Beam, R. D. (2012). The collaborative forest landscape restoration programme: A history and overview of the first projects. *Journal of Forestry, 110*, 381–391.

Semwal, R. L., Nautiyal, S., Sen, K. K., Rana, U., Maikhuri, R. K., Rao, K. S., & Saxena, K. G. (2004). Patterns and ecological implications of agricultural land-use changes: A case study from central Himalaya, India. *Agriculture, Ecosystems and Environment, 102*, 81–92.

Semwal, R. L., Nautiyal, S., Maikhuri, R. K., Rao, K. S., & Saxena, K. G. (2013). Growth and carbon stocks of multipurpose tree species plantations in degraded lands in Central Himalaya, India. *Forest Ecology and Management, 310*, 450–459.

Sharma, C. M., Baduni, N. P., Gairola, S., Ghildiyal, S. K., & Suyal, S. (2010). Tree diversity and carbon stocks of some major forest types of Garhwal Himalaya, India. *Forest Ecology and Management, 260*, 2170–2179.

Stanturf, J. A., Palik, B. J., & Dumroese, R. K. (2014). Contemporary forest restoration: A review of emphasizing function. *Forest Ecology and Management, 331*, 292–323.

Takeuchi, K., Elmqvist, T., Hatakeyama, M., Kauffman, J., Turner, N., & Zhou, D. (2014). Using sustainability science to analyse socio-ecological restoration in NE Japan after the great earthquake and tsunami of 2011. *Sustainability Science, 9*, 513–526.

Vanderhaegen, K., Verbist, B., Hundera, K., & Muys, B. (2015). REALU vs REDD+: Carbon and biodiversity in the Afromontane landscapes of SW Ethiopia. *Forest Ecology and Management, 343*, 22–33.

Wakeel, A., Rao, K. S., Maikhuri, R. K., & Saxena, K. G. (2005). Forest management and land use/cover changes in a typical micro watershed in the mid elevation zone of Central Himalaya, India. *Forest Ecology and Management, 213*, 229–242.

Wheeler, C. E., Omeja, P. A., Chapman, C. A., Glipin, M., Tumwesigye, C., & Lewis, S. L. (2016). Carbon sequestration and biodiversity following 18 years of active tropical forest restoration. *Forest Ecology and Management, 373*, 44–55.

Yami, M., Mekuria, W., & Hauser, M. (2013). The effectiveness of village bylaws in sustainable management of community-managed exclosures in Northern Ethiopia. *Sustainability Science, 8*, 73–86.

Chapter 5
Social-Ecological Transformation Through Planting Mixed Tree Species on Abandoned Agricultural Land in the Hills of Nepal

Bishnu Hari Pandit, Netra Kumari Aryal, and Hans-Peter Schmidt

Abstract A project entitled, "Building village economies through climate farming & forest gardening" (BeChange) was implemented in four municipality areas of the Tanahun and Lamjung districts of Nepal from May 2015. In order to assess changes in the social-ecological system that result from this project targeting abandoned agricultural lands, this case study was conducted using various methods: triad grouping, GPS point surveys, household surveys, focus group discussions (FGDs), field observation and reports. A participatory approach in reforestation on abandoned agricultural land with introduction of carbon credits has become a new livelihood strategy for local communities. It has not only attracted domestic and international tourists, but also helped to conserve biodiversity and local ecology. This activity also united village women and indigenous communities as triad groups for collaborative outcomes. A total of 42,138 seedlings of mixed tree species such as *Michelia champaca*, *Elaeocarpus ganitrus*, *Bassia butyraceae*, *Bauhinia purpurea*, and *Cinnamon tamala* were planted by 276 families on abandoned agricultural land between May 2015 and July 2018. However, as of 2020, this range has expanded to include 635 families with plantations of more than 65,000 seedlings. The set-up and maintenance of these forest gardens were financed with advanced payments for the carbon sink services of the planted trees. Farmers who succeeded with tree survival rates above 80% received an additional yearly carbon sink payment. The outcomes of the project show significant improvements in food security and tree biodiversity in the project villages. Of the total sampled households, almost half (45%) were under extreme poverty and had food sufficiency for only 3 months/year before the project.

B. H. Pandit (✉)
Kathmandu Forestry College, Koteshwor, Kathmandu, Nepal

Ithaka Institute for Climate Farming, Tanahu, Nepal
e-mail: bishnu@kafcol.edu.np

N. Kumari Aryal
Krishiban Prabhardan Nepal, Koteshwor, Kathmandu, Nepal

H.-P. Schmidt
Ithaka Institute Ancienne Eglise 9, Arbaz, Switzerland

© The Author(s) 2021 77
M. Nishi et al. (eds.), *Fostering Transformative Change for Sustainability in the Context of Socio-Ecological Production Landscapes and Seascapes (SEPLS)*,
https://doi.org/10.1007/978-981-33-6761-6_5

With the project, this percentage dropped to 22%, signals the emergence of seeds for transformative change.

Keywords Ecological transformation · Carbon sink · Forest garden · CO_2 certificate · Food security · Abandoned agricultural land · Triad group

5.1 Introduction

Nepalese people have practiced organic agriculture for hundreds of years across the country's hills. Terraced slopes and water channels were made manually. Farmers allowed trees to grow on the terrace faces and edges, in a system that has been well described in many studies (Malla 2000; Regmi and Garforth 2010; Pandit et al. 2018). The typical Nepali trees growing in farmlands were mainly fodder trees for livestock. The resulting manure and forest litter were then used as mulching material to maintain soil fertility. Farmers cultivated rice, maize, wheat, millet and vegetable crops on their farms that were commonly terraced and bounded with trees. However, these terraces have been largely abandoned over the last decade. Studies indicate that on average 37% of arable land is abandoned in Nepal (Paudel et al. 2014; Ojha et al. 2017). Increasing land abandonment in Nepal poses multiple threats related to food insecurity, loss of rural livelihoods, reduction in crop production, loss of biodiversity and soil productivity, and damages to the ecological landscape. The implications of agricultural land abandonment are not limited only to the household level, but also extend to impacts on the national economy. For example, the GDP contribution of the agricultural sector was 33% in 2011, though this figure decreased to 26% in 2018 (CBS 2018), showing agricultural land abandonment as a major problem for the people, economy, and environment (Basnet 2016; Ojha et al. 2017). As a result, 42 of the total 75 districts are reported to be food insecure (FAO 2010). This indicates that there is significant room for improvement in the contribution of tree growing on such land (Schmidt et al. 2017). Furthermore, it is important to create jobs in rural areas in Nepal in the wake of COVID-19, as migrant youths are returning and searching for new jobs.

In recent decades, terms such as social-economic transition, societal transformation, ecological transformation, green economy and sociotechnical transition have increasingly been discussed (McDowell 2012; Brand and Wissen 2017). Social-ecological transformation is an umbrella term which describes political, socioeconomic, and cultural shifts resulting from attempts to address the socioecological crisis (Brand and Wissen 2017). Social-ecological transformation is a systemic approach applied to broad-based change in social-ecological systems that catalyses rapid shifts in the mental constructs inhibiting solutions to complex problems of the socio-ecological landscape that prevent it from realising its full potential (Walker et al. 2004; Brand and Wissen 2017). There is an urgent need to change our society, particularly because of impending and potentially catastrophic climate disruption and degradation of ecological life-support systems (Butzer 2012; Pearson and

Pearson 2012). In addition, transformation requires proactively changing structures and processes when conditions make the existing social-ecological system untenable (Butzer 2012; Pearson and Pearson 2012).

The life-support systems in the landscape of this study area have been affected adversely, hindering people's ability to sustain their livelihoods and exacerbating the poverty level. A fundamental shift in human behavior is required, to live more ethically and efficiently and to radically rethink the concept of progress and economic development in our societies. We are all part of one planet; our wellbeing depends on working together for a sustainable, more equitable society. Transformational change is a formidable challenge but is necessary. In an effort to improve understanding of socio-ecological transformation, this study examines whether a project conducted in a rural area of Nepal has the potential to become the seeds of transformative change. The "Building village economies through climate farming & forest gardening" (BeChange) project planted trees on abandoned farmlands to improve environmental and livelihood benefits. This project could be assessed for its potential to serve as the seeds of social-ecological transformation.

5.2 Methodology

5.2.1 Study Site

The study site is located in middle hill region of Nepal (Fig. 5.1). Four municipalities, two each from Tanahun and Lamjung Districts within Gandaki Province, were selected before project intervention. From each of the four municipalities, one focus village was selected purposively for this case study. The total area of the four municipalities is 47,200 ha, and the population is 103,680 (Table 5.1). This area is connected by the Prithibi Highway, some 130 km away from Kathmandu, the capital city of Nepal. Of the four municipalities, Bhanu is situated in between the other three (Bandipur, Rainas and Madya Nepal) (Fig. 5.1). The geographic coordinates are 28° 1′ 48.0004″ N, 84° 26′ 24.0072″ E, and the altitude of the study area ranges from 418 m to 2320 m above mean sea level. The climate is of the humid subtropical type across all four municipalities. The mean temperature at the lower hills is 16 °C. The winter is cool and dry.

Rainfall data shows that the area receives more than 80% of its annual rain within the short period of 15 May to 15 September. The monsoon starts in the beginning of May and continues until September, while the remaining months remain generally dry. The average annual rainfall ranges from 2000 to 2500 mm. There is general decline in soil quality with rising elevation. The soils of the upper elevation are medium to light-textured, highly permeable, acidic and of low to medium fertility. The hill slopes tend to lose their top-soils, owing to erosion caused by their steepness and periods of intensive rainfall. The soils of the lower elevation are of better quality and fine textured.

Figure 1:Map of Project Area

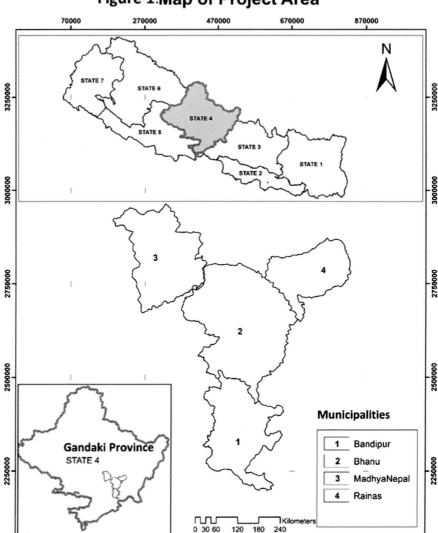

Fig. 5.1 Map of project area. The location of the study area is also shared at the google map link below. https://www.google.com/maps/d/edit?mid=10dFDhfUPe3wOIwHAuVcpZpHcjVg-ipWD&usp=sharing

The vegetation is characterised by open, mixed *Shorea* and *Schima* forests. *Shorea* (commonly known as sal trees) forest is found along the foothills and lower parts of the hill slopes. *Schima wallichii*, *Castanopsis indica*, and *Eugenia jambolana* are the typical tree species found in forests. *Emblica officinalis* and *Terminalia* species are the dominant tree-based non-timber forest product (NTFP)

Table 5.1   Basic information of the study area

Features of study area	Description
Province	Gandaki
District	Tanahun and Lamjung
Municipalities	Bandipur, Bhanu, Madya Nepal, Rainas
Size of geographical area (hectare)	47,200
Number of indirect beneficiaries	103,680
Size of the case study/project area (hectare)	42
Number of direct beneficiaries	276
Number of respondents	121
Dominant ethnicity(ies), if appropriate	Gurung, Magar and Chhetri/Brahmin
Geographic coordinates (latitude, longitude)	28° 1′ 48.00″ N; 84° 26′ 24.00″ E

species, among others such as *Bassia butyraceae*, *Myrica esculenta*, *Terminalia balerica*, *Eugenia jambolana* and *Dandrocalamaus strictus*, that are generally grown on farmland terraces. Rice is grown on the terraces of lower hills, and while maize and millet are cultivated in the uplands. These are supplemented by fields grown with vegetables and good quality upland crop terraces. Farmers have been increasingly compelled to keep their land fallow or abandoned in this area, as the current level of returns from cropping does not meet the cost of cultivation. Of the total agricultural land, farmers have kept more than half of their land fallow. This proportion is even higher in the areas where tree planting is not carried out. It is for this reason that the forest gardening project was initiated.

5.2.2   Methods

To assess the impact of the project, the following methods were applied throughout the project: process documentation by forming triad group, GPS mapping, household surveys, focus group discussions (FGDs) and examining the results of carbon accounting through tree planting.

5.2.2.1   Establishment of a Triad System and Carbon Crediting

The project applied a triad system in which three families formed a group. Each family in the group was responsible for planting and growing trees on their land. Families communicated and shared the recurrent tasks amongst themselves, frequently patrolling the plantations of each family, determining failure rates, and replanting together. Each member of the triad was responsible for the other two partners as well. Only when all three families could show tree survival rates of 80% or higher, were they entitled to receive the carbon premium. If, for example, one family reached 97%, another 83%, but the third only a 72% tree survival rate, then the triad as a whole failed and none of the families earned the premium. If all three partners within the triad surpassed the 80% mark, each one of them received the carbon bonus in accordance with the number of trees grown on their own terraces. Triads that did not reach the 80% mark had to pay for new seeds themselves, but after successful replanting were paid the carbon premium in the total amount of 60 NPR/tree (digging pit = 20 NPR + irrigation and management = 20 NPR + final payment = 20 NPR) for the number of trees that survived.

At least 15 tree species were planted based on farmers' preferences. These include: *Artocarpus lakoocha, Morus alba, Melia azedarach, Bassia butyraceae, Michelia champaca, Emblica officinalis, Cinnamomom tamala, Choerospondias axillaris, Terminalia bellirica* or *chebula, Paulownia tomentosa, Citrus limon, Leucaena leucocephala, Flemingia congesta, Pinus* spp. and *Sapindus mukorossi.* Of the total species, two-thirds (10 out of 15) were local and indigenous. Biochar-based fertilisers (1 part feedstock—*Euphatorium* biomass: 1 part urine: 4 part cow-dung mix) were applied to these plantations. The results of biochar application were highly effective.

5.2.2.2   GPS Mapping and Tree Growth Measurement

Every certified tree was accurately mapped and dated with a GPS-based smartphone application, and plot-wise mapping was also performed. In each plot, the number of trees planted was recorded and verified. At the end of the first (2015/16) and second year (2016/17), tree height and trunk diameter at 10 cm above ground were measured. After the third year, trunk diameter was measured at breast height (150 cm above the ground), the tree's general vigour and health were rated on a scale from 1 to 10, and a picture of each tree was taken. This monitoring system has served to ensure that CO_2 certificates were issued only for trees that were actually growing well. Customers who bought CO_2 certificates issued by the Ithaka Institute for Carbon Strategies could then know where "their" trees were, how well they were growing, and how the biomass carbon was eventually sequestered. The goal is that CO_2-subscribers based in Europe (Example: Eubenheimer Manufaktur UG, Email: info@eussenheimer-manufaktur.de) are able to follow the growth of the forest garden online in order to reclaim their CO_2 emissions and to know which family does the work for them, forming a virtual alliance.

5.2.2.3   Survey on Food Security in Households

In order to assess the impacts of the forest gardening project on food security and poverty alleviation, a sample of 121 household heads among a total of 276 were interviewed. These were households that had planted trees during the year 2018/19. However, the total households that planted trees as of the end of 2019 scaled out to 635. A range of 40–60% of households were sampled from the four villages (Table 5.2). This survey mainly identified the priority activities of farmers across

Table 5.2 Sampling of households

Village/ municipality	Brahmin/ Chhetri	Indigenous (Gurung, Magar and Tamang)	Dalit (shoe maker, iron smith and teller)	Total HHs	Sample HHs
1. Ratanpur, Bhanu Municipality	89	38	22	149	59 (40%)
2. Bandipur Rural Municipality	2	82	2	86	39 (45%)
3. Rainas Municipality	20	4	1	25	13 (50%)
4. Madya Nepal Municipality	13	2	1	16	10 (60%)
Total	124	126	26	276	121

each of the ethnic groups. This was helpful to investigate which households suffered food insufficiency and had fallen into a poverty trap at the end of the project intervention.

5.2.2.4 Focus Group Discussions (FGDs)

In each of the four target villages (Fig. 5.1), at least one FGD was conducted. FGDs helped in classification of households in terms of their focused activities such as (1) gender sensitisation and tree planting for increased productivity and carbon sequestration, and (2) increased sustainable income through agroforestry, ecotourism for food security and poverty reduction.

5.2.2.5 Setting Indicators for Assessing Social-Ecological Transformation

As per the project goal, four objectives and 11 key indicators were set for measuring the success of the project (Table 5.3). These indicators were assessed with different tools, mainly household surveys and FGDs, discussed above.

5.3 Results

5.3.1 Tree Planting by Triad Family Groups

One year after the initial forest gardens were planted, the results were quite variable. While some families had kept 95% of their trees alive, less than 30% trees survived

Table 5.3 Indicators for success (2015/16 to 2018/19)

Objectives	Objectively verifiable indicators
1. Increase water availability and water retention capacity	1. At least 75 households at both upstream and downstream watershed have secured drinking water source
	2. At least 25% of participating farmers have access to irrigation water in dry season
	3. Agricultural productivity increase of 25% for all participating farmers after application of biochar-based organic fertiliser
2. Improved biodiversity and ground cover change	4. At least 22 ha of abandoned agricultural land of 100 households covered by multiple tree species plantation
	5. Plantation area increased by 50% at the end of the project period
	6. Participating farmers earn additional income from sales of agricultural products from abandoned land
3. Increased productivity of cash crops income and food security for the poor	7. Income of at least 20 poor and socially excluded families (more than 50% women) in upstream and downstream areas increased by 25%
	8. Cash crops (vegetables and spices) increased by 50% for 10 water users in lower hills and 15 poor and socially excluded families
	9. Food security level of participating farmers increased by 25% at the end of the project
4. Policy and institutional transformation	10. Policy feedback available for change in the private forest rules and regulations/carbon payment at both local and national levels
	11. Incentive mechanism created for trees planted on abandoned agricultural lands at local level

for other families. The average survival rate was only 60% during the first year (Table 5.4). This year was the most challenging, and farmers with low survival rates replanted the trees on their own to compensate the carbon sequestration rates for payment. Most plantings reached success rates of 60–70%. Although this figure is more than twice the rate of other reforestation projects in Nepal, there was certainly much room for improvement.

Since the summer of 2015, a total of 42,502 trees have been planted in cooperation with 276 farming families (Table 5.4); 49 ponds have been put into place; and four villages are now housing tree nurseries. While only 60% of the planted trees survived the first year, with 60 tons of CO_2 extracted from the atmosphere, the survival rate for the second and third years reached 70.25% and 77.5%, respectively. The introduction of triads improved the success of the plantings. However, this system worked well only when the community households were located close to each other.

Table 5.4 Number of villages, beneficiary households and total trees planted and survived by year

Village/ municipality	First two-year period				Second two-year period			
	2015/16		2016/17		2017/18		2018/19	
	No. of villages	No. of HHs	No. of villages	No. of HHs	No. of villages	No. of HHs	No. of villages	No. of HHs
Ratanpur, Bhanu	1	40	2	66	8	143	2	149
Bandipur	0	0	1	20	1	75	1	86
Madya Nepal	0	0	0	0	0	0	1	25
Rainas	0	0	0	0	0	0	1	16
Total	1	40	3	86	9	218	5	276
Trees planted (survival rate)	10,442 (60%)		9838 (77.5%)		11,946 (87.6%)		10,276 (53%)	

5.3.2 Carbon Credits

In the carbon calculation, we used a wood density database reference from the google sheet for each of the species planted. When farmer families in Nepal plant 583 trees on a hectare of abandoned rice terraces, CO_2 removal would be 336 t over 20 years. For example, over a period of 20 years, a cinnamon tree extracts 70 kg, a *Michelia champaca* 380 kg, and a frequently coppiced *Melia* tree 557 kg (Table 5.5). A bio-diverse mix of 583 trees per hectare results in 336 t CO_2 being pulled from the atmosphere over the course of 20 years. Therefore, one tree sequesters CO_2 equivalent to 576 kg (336 t/583 trees). Besides the carbon sequestered in trees, about 80% of the biochar carbon applied to the soil is stable for a period of hundreds of years (Lehmann et al. 2015; Zimmerman and Gao 2013). The invasive species *Eupatorium odoratum*, locally called *banmara* (meaning "forest killer plant"), was used as a main feedstock for making biochar. This species is abundantly available in forests and has a high regeneration capacity. Therefore, there is no problem with the ecological integrity of use of *Eupatorium* feedstock. The application of biochar to agricultural crops has proven to be very useful for enhancing crop productivity. Examples from the project site demonstrated a fourfold increase in pumpkin yield and doubled the yield of cabbage and cauliflower inside forest gardens. On an average there has been at least a 25% increase in crop yield. Carbon sequestration potential has also been increased by 50% in trees planted with the use of biochar, which is not currently accounted for in carbon payments.

The carbon calculation here is based on seven out of 58 important tree species from the catalogue of carbon fixation by trees (Table 5.5). The rate of carbon per ton sold in Europe was 35 EUR and in the United States was 12 USD in volunteer markets. This covers the costs of seedlings, digging pits, planting, irrigation and management. On top of this, at the end of each of the 3-year periods, farmers were paid 20 NPR per survived tree as a reward. After 3–4 years, farmers were supposed

Table 5.5 Carbon fixation by trees planted in Bandipur hills for 20 years

Tree species	Stem volume (πr2*Ht*.5)/4	Wood density kg/m³	Trunk biomass	Branch biomass (44%)	Root biomass (12%)	Leaf biomass (12.5%)	Total biomass over 20 year/tree (kg)	Total C/Tree (kg)	CO2 removal/tree (kg)	No of tree/ha	Total C t/ha
Pinus roxburgii	0.707	327	231	102	28	29	389	183	671	400	268
Michelia champaca	1.202	400	481	212	58	60	810	381	1396	278	388
Melia azedarach	1.256	560	703	309	84	88	1185	557	2042	400	817
Cinnamomum tamala	0.188	510	96	42	12	0.0	150	70	258	625	161
Emblica officinalis	0.097	680	66	29	08	08	111	52	191	625	120
Ficus benjamina	0.721	490	353	155	42	44	595	279	1026	156	160
Saraca ashoka	0.188	496	93	41	11	12	157	74	271	1600	434
									Total (avg)	583	336

to earn from the sale of tree products and other associated activities, such as selling honey and earning from homestays and ecotourism.

5.3.3 Impacts on Food Security of Households

Of the total 65 indigenous (Gharti, Gurung/Magar) and Dalit households sampled, almost all, or 54 households (83%), had food sufficiency for less than 3 months/year prior to the start of the project (Table 5.6). This figure dropped to 34 households as of the end of the project, of which 14 households shifted to 4–7 months and six households to 8–12 months food sufficiency level. If the impact on Dalit families is looked at alone, of the total 18 Dalit families that were under extreme poverty (food sufficiency for only 3 months), six households shifted to the next level (4–7 months), and two households shifted to the third level (8–12 months) (Table 5.6). These improvements are attributed mainly to goat and fodder tree support, ecotourism and homestay programmes.

5.3.4 Level of Impact as Shown by Changes in Indicators

Assessment of progress on the 11 indicators showed significant impacts except for Indicator 2 (Table 5.7). The inadequate availability of irrigation water to collect from water sources in winter is thought to be the cause for lack of progress on this indicator. An investment in technology that captures rain water in the winter is required. To address this issue, a dew collection net was set up in one of the villages, but this method also did not perform well. While every household now has safe and secure drinking water, they do not have enough water for irrigation in both the upper

Table 5.6 Outcomes achieved through implementation of priority activities by caste and ethnicity

| Activity | Ethnicity | Food sufficiency in households (HHs) | | | | | |
| | | Up to 3 months | | 4–7 months | | 8–12 months | |
		Pre-project	Post-project	Pre-project	Post-project	Pre-project	Post-project
Gender sensitisation and tree planting for increased productivity and carbon sequestration	Brahmin	24	19	9	4	5	1
	Chhetri	10	7	4	2	4	1
	Sub-total	34	26	13	6	9	2
Increased sustainable income through agroforestry and ecotourism for food security and poverty reduction	Gharti	7	3	1	3	1	1
	Gurung/ Magar	29	21	4	5	2	3
	Dalit	18	10	2	6	1	2
	Sub-total	54	34	7	14	4	6
Total		88	60	20	20	13	8

Table 5.7 Progress on indicators

Indicators set before project (2015)	Present level of progress on indicators (2020)
1. At least 75 households at both upstream and downstream watershed have secured drinking water	A total of 79 households have secured drinking water
2. At least 25% of participating farmers will have access to irrigation water in dry season	Only 11% of farmers have access to irrigation in dry season
3. Agricultural productivity increase of 25% due to application of biochar	Agricultural productivity increased by more than 25%
4. At least 22 ha of abandoned agricultural land of 100 households covered by multiple tree species	68 ha of land belonging to more than 200 families was converted into forest gardens and three new indigenous species regenerated
5. Plantation area increased by 50% at the end of project	Plantation area increased triple-fold
6. Participating farmers earn additional income	Increase in additional income (Table 5.5)
7. Income of at least 20 poor and socially excluded families (more than 50% women) increased by 25%	20 poor and Dalit families shifted from low food security level to high security level (Table 5.6)
8. Cash crops (vegetables and ginger) increased by 50% for 10 water users in lower hills and 15 poor and socially excluded families	All targeted (10+15) farmers have cultivated vegetable and cash crops in winter
9. Food security level of participating farmers increased by 25% at the end of the project	32% of households shifted from low food security to high food security level
10. Policy feedback available for change in the private forest rules and regulations/carbon payment at both local and national levels	Carbon payment is provisioned in the new Forest Act-2019
11. Incentive mechanism created for trees planted on abandoned agricultural lands at local level	Municipality created funds to support tree planting on abandoned agricultural land

and lower watershed of the landscape. Because of biochar application, the fertility level of soil is believed to have increased by 10%, and the impact of which has been food sufficiency level also increasing by 31%. Plant diversity level has also increased significantly. Some indigenous species such as *Schima wallichii*, *Syzygium cumini*, *Ficus religiosa*, and *Ficus bengalensis*, which were not seen before, have now started to regenerate themselves inside forest gardens. The income level of participating farmers also increased and positive policy feedback has been evident. The local government and divisional forest office are now actively involved in rehabilitation of the landscape and co-funding.

5.3.5 Stakeholders' Roles in Achieving Multiple Benefits from SEPLS

This project has identified the various stakeholders directly or indirectly involved in providing multiple benefits to the SEPLS. The most important stakeholder group encompasses local farmers including women, indigenous people and Dalit (Indicator 7), who were actively involved in planting trees on their land in small groups, called triad groups, and also benefited from the activities. The community forestry user group (CFUG) also provided assistance to its members through promotion of tree nurseries and tree planting activities (Indicator 4). A local implementation committee was formed for monitoring and evaluation of the success of the project. This committee included a representative from the local school, a member from local government (municipality), a women's group leader, a retired school teacher and the CFUG committee chair. Actors with policy knowledge and values involved the divisional forest office (DFO) staff, the agriculture knowledge centre and the municipality head. DFO and the Municipality jointly cofounded this initiative (Indicator 11). Staff from the Forest Research and Training Centre (FRTC) under the Ministry of Forests and Environment (MOFE) provided time-to-time follow up and technical support in assessment of the effects of insect pests and diseases emerging in the forest garden. All of their actions and cooperation helped the project to ensure multiple benefits for the landscapes.

5.3.6 Scaling Out and Scaling Up

To address climate related risks and biodiversity loss, and to enhance the livelihoods of people living in the landscape, this project has already involved 276 households in the planting of more than 40,000 trees on abandoned agricultural land. We observed that many other local and indigenous tree species have regenerated inside the forest gardens. This project has been scaled out in two more communities in two other districts in the province (Pokhara and Gorkha). Pame village of Ward 24 of Pokhara Municipality has initiated a programme with at least 20 farmers and one farmer-managed commercial nursery. Activities have been continued by the local communities even after the project ended, and the number of forest garden promoters has reached 635, with 65,000 seedlings planted on their farms as of 2020.

5.4   Discussion

5.4.1   Success of Tree Planting and Transformation

A good number of trees from the first plantation campaign have already grown so much that some species (*Cinnamomum tamala* and *Morus alba*) are now at the harvesting stage. The visible excitement of the farmers over witnessing the fruits of their labour and how the landscape has changed over such a short time span is a great success of this project, and also serves to motivate farmers toward new steps and trials to promulgate our findings. Over time, the outcomes and successes (Table 5.7) can be used to attract migrant youths to return. Mainly two aspects of the project have motivated the communities to continue the forest gardening activities on their land. The first is project assistance to protect drinking water sources and distribution systems. The second is payment for carbon credits. Likewise, at present homestays and ecotourism are associated with the motivation of local people to make these changes.

5.4.2   Impact of Carbon Credits

Carbon credits and carbon markets are a component of national and international attempts to mitigate the rise in concentrations of greenhouse gases in the atmosphere. A carbon credit (carbon offset) is a credit for greenhouse emissions reduced or removed from the atmosphere by an emission reduction project, which can be used by governments, industry, or private individuals to compensate for the emissions they generate elsewhere. Offsetting one metric ton of carbon means that there will be one less Mt of carbon dioxide in the atmosphere. In our case, one tree sequesters an average 28.8 kg of CO_2 eq per year (Table 5.5), which means 42,000 trees have sequestered more than 1200 tons of carbon annually. The payments received by farmers cover the cost of seedlings, pitting, and management, as well as 20% as a reward to the farmers at the end of each year. Besides these economic benefits, farmers get benefits from tree products: fruits, nuts, medicines, essential oils, silk, perfume, honey, timber and animal fodder, including other added values such as organic matter increase, biodiversity, erosion control, and water retention.

5.4.3   Opportunities for Increasing Income Through Value-Added Products

In the southern lowlands of Nepal bordering India, fertilisers, pesticides and machines are cheaper, and market access is easy. By contrast, in the poorly accessible hills of Nepal, where no fertilisers and pesticides can be purchased and most of the soil is degraded, biochar-based fertiliser can be helpful. On the steep terraces, it is not profitable to grow grain beyond that needed for personal nutrition. The villages are too far away from the marketplaces, the roads are bad, and it is nearly impossible to mechanise production. All of these impediments can be overcome if higher-value crops are planted and processed on-site. With local value addition, the durability of the goods increases, the transport volume decreases, and marketing becomes economically viable. In this manner, silk or tea from mulberry and moringa leaves can be produced and sold instead of rice; essential cinnamon oil instead of corn; nuts instead of potatoes; and dried banana or papaya chips instead of millet, providing direct income through value addition.

However, there are other income sources which have indirectly benefited participant farmers, such as cultural ecotourism and biochar fertilisation. The inclusion of a cultural tourism component in the tree planting project provided another innovative solution to diversify the income of the villagers while creating seasonal jobs for the village youth. It also offered opportunities for the villagers to learn new skills and motivated them to gain specialised education related to tourism and hospitality. The biochar fertilisation on the other hand has increased the fertility level of soil, which in turn increased the yield and biomass of both crops and trees that are being sold. The use of biochar fertiliser not only increases the crop yield, but also provides an alternative to scarce and costly chemical fertiliser that is imported from India.

5.4.4 Project Scaling Out and Up Is Possible

As a scaling up activity, this project secured co-funding from the Bhanu Municipality to pay for nursery operator at Ratanpur. Similarly co-funding from Ithaka Foundation Switzerland helped to develop a soil organic matter assessment laboratory at Ratanpur, Tanhu, and a knowledge hub centre at Satungal, Kathmandu. Over the last 2-year period (2019–2020), this project has been scaling out its activities to more than 635 households in five to six local government areas of four districts and has planted an additional 200,000 seedlings. This is a very important innovative activity to motivate farmers to proactively participate and expand the volume and quality of activities.

5.5 Lessons Learned and Conclusions

5.5.1 Lessons Learned

- The established forest gardens have created a new environmental balance by bringing back trees on degraded and barren land.
- Tree planting on abandoned agricultural land has been proven successful in reducing the concentration of atmospheric carbon dioxide, a leading cause of

global climate change, and at the same time providing economic benefits to the farmers.

- This type of project can provide a diverse array of value-added products that have high market potential.
- The use of biochar-based fertilisers for tree planting boosted the growth of both the trees and the crops cultivated between and under the trees, and could replace expensive chemical fertiliser.
- The inclusion of ecotourism as a component of the project has helped to diversify the income of the villages. The forest gardening project can be further scaled out in the context of the post-Covid-19 situation.

5.5.2 Conclusions

With the development of the forest garden system, this project has been successful in motivating a large number of farmers to undertake activities to sustain their livelihoods. Income from the sale of carbon credits is enough to establish forest gardens and to get additional income to buy food. The creation of value-added products such as essential oils, biochar, fruits, nuts and other agricultural products would be very effective to further promote tree planting on the degraded hills of Nepal. Working on value chains with renewable energy, such as post-harvest value-adding of forest products like drying and distillation, would help to increase the shelf-life and decrease the cost of transport. These value-adding processes, which include indirect drying of fruits by hot air in solar dryers that retain nutrients, make forest products lighter and of lesser volume to transport. Heat recovery devices used for essential oil production while making biochar have been very effective, which could also be promoted at a wider level. In the wake of Covid-19, many Nepali people are migrating back home and mass unemployment has already begun. This type of project could help these people to develop sustainable agroforestry practices on abandoned agricultural land and lead the path towards social-ecological transformation.

References

Basnet, J. (2016). *CSO Land reform monitoring report*. Kathmandu, Nepal: Community Self Reliance Centre.

Brand, U., & Wissen, M. (2017). Social-ecological transformation. In D. Richardson, N. Castree, M. F. Goodchild, A. Kobayashi, W. Liu, & R. A. Marston (Eds.), *International encyclopedia of geography: People, the earth, environment and technology* (pp. 1–21). Hoboken, NJ: Wiley.

Butzer, K. W. (2012). Collapse, environment, and society. *Proceedings of the National Academy of Sciences, 109*, 3632–3639.

CBS. (2018). *Agriculture statistics*. Thapathali, Kathmandu, Nepal: Department of Central Bureau of Statistics.

FAO. (2010). *Assessment of food security and nutrition situation in Nepal (An input for the preparation of NMTPF for FAO in Nepal)*. Pulchowk, Nepal: Food and Agriculture Organization of the United Nations.

Lehmann, J., Abiven, S., Kleber, M., Pan, G., Singh, B. P., Sohi, S. P., & Zimmerman, A. R. (2015). Persistence of biochar in soil. In J. Lehmann & S. D. Joseph (Eds.), *Biochar for environmental management* (pp. 235–299). London: Routledge.

Malla, Y. (2000). Farmers' tree management strategies in a changing rural economy, and factors influencing decision on tree growing in Nepal. *International Tree Crops Journal, 10*, 247–266.

McDowell, G. (2012). The role of bridging organizations in facilitating socio-ecological transformation: A case study of the Great Northern Landscape Conservation Cooperative. MSc thesis, Environment Change Institute, University of Oxford, UK.

Ojha, H. R., Shrestha, K. K., Subedi, Y. R., Shah, R., Nuberg, I., Heyojoo, B., Cedamon, E., Rigg, J., Tamang, S., Paudel, K. P., Malla, Y., & McManus, P. (2017). Agricultural land underutilisation in the hills of Nepal: Investigating socio-environmental pathways of change. *Journal of Rural Studies, 53*, 156–172.

Pandit, B. H., Nuberg, I., Shrestha, K. K., Cedamon, E., Amatya, S. M., Dhakal, B., & Neupane, R. P. (2018). Impacts of market-oriented agroforestry on farm income and food security: insights from Kavre and Lamjung districts of Nepal. *Agroforestry Systems, 93*, 1593–1604.

Paudel, K. P., Tamang, S., & Shrestha, K. K. (2014). Transforming land and livelihood: analysis of agricultural land abandonment in the mid hills of Nepal. *Journal of Forest and Livelihood, 12* (1), 11–19.

Pearson, L. J., & Pearson, C. J. (2012). Societal collapse or transformation, and resilience. *Proceedings of the National Academy of Sciences, 109*, E2030–E2031.

Regmi, B., & Garforth, C. (2010). Trees outside forests and rural livelihoods: a study of Chitwan District, Nepal. *Agroforestry Systems, 79*, 393–407.

Schmidt, H. P., Pandit, B. H., Kammann, C., & Taylor, P. (2017). Forest gardens for closing the global carbon cycle. *The Biochar Journal, 2017*, 48–62.

Walker, B., Holling, C. S., Carpenter, S. R., & Kinzig, A. (2004). Resilience, adaptability and transformability in social-ecological systems. *Ecology and Society, 9*(2), 5.

Zimmerman, A. R., & Gao, B. (2013). The stability of biochar in the environment. In N. Ladygina & F. Rineau (Eds.), *Biochar and soil biota* (pp. 1–40). Boca Raton: CRC Press.

Chapter 6
Transformative Change Through Ecological Consumption and Production of Ancient Wheat Varieties in Tuscany, Italy

Guido Gualandi and D. Williams-Gualandi

Abstract In the 2016 Volume 2 of the Satoyama Initiative Thematic Review (SITR), a summary of the activities of the Grani Antichi Association in Montespertoli, Tuscany, was introduced with a roll out plan for transformative change of the supply chain and possible replication of the project in other regions. The main goal of the project has been to restore and preserve ancient varieties of wheat, cultivate them sustainably and include a form of payment for the least compensated members of the production chain. The aims of the Association are to reduce the carbon footprint of modern agricultural practices and the landslides and soil erosion caused by them, to preserve biodiversity and most importantly, to improve farmers' revenue, enabling them to safeguard the environment and improve health by cultivating healthy food. The preservation of social ties and local knowledge is an additional result. Markers of the project's success include benefits that are equally distributed across the production chain, farmers who are motivated to cultivate ancient wheat varieties and the conversion of 500 ha of abandoned or conventionally cultivated land to a more sustainable and biodiverse system. The market economy system in place was dismantled, and farmers now have access to more economic benefits, which must be distributed fairly. Because the project provides a transformative model of production and consumption outside the traditional market economy system, it appears to function with a complete multi-sectoral chain, where producers, food processors and consumers agree on a set price for a defined product. This chapter provides a preliminary analysis of the successes and challenges related to the main project and to upscaling in different areas.

Keywords Grain · Biodiversity · Sustainable farming · Scale up · Farmers' cooperative

G. Gualandi (✉)
Associazione Grani Antichi di Montespertoli, Tuscany, Italy
e-mail: guido@guidogualandi.com

D. Williams-Gualandi
NHL Stenden University of Applied Sciences, Leeuwarden, Netherlands
e-mail: debra.williams.gualandi@nhlstenden.com

M. Nishi et al. (eds.), *Fostering Transformative Change for Sustainability in the Context of Socio-Ecological Production Landscapes and Seascapes (SEPLS)*,
https://doi.org/10.1007/978-981-33-6761-6_6

6.1 Introduction

Biodiversity loss triggered by modern agriculture is a serious concern for local governments, research centres, producers and consumers the world over. In Italy, some regional governments have sponsored activities involving local flora and traditional foods, including the preservation of edible plants. The region of Tuscany has been particularly active in preserving a number of crop and livestock varieties to ensure the continuation of specific species. It is known that some older and more resilient plant varieties are more adaptive to climate change impacts, such as higher temperatures and the less rainfall, as well as soil erosion. For this reason, these varieties are considered socially and environmentally beneficial. However, the trend to encourage cultivation of older grain varieties has been slow to develop because grain revenues are very low and governments do not provide incentives to grain farmers. This chapter focuses on a project involving wheat.

In the specific case of wheat (*Triticum*), the process of selection and hybridisation, which has taken place over a span of more than 10,000 years, resulted in an abundance of wheat varieties throughout the world. Each variety adapted to the growing conditions of its local environment. However, genetic engineering and the processes of agricultural industrialisation have radically decreased wheat genotypes. Until quite recently, farmers could only choose from a few modern varieties available for purchase. As the national registry for plant species did not include ancient varieties, it had become more and more difficult to cultivate those varieties legally. Although modern varieties are attractive for their high yield, they require the use of fungicides, pesticides and fertilisers, all of which compromise the health of the ecosystem. Local ancient varieties offer many potential benefits as they are healthier for human consumption, more sustainable and adaptable to their environment and require no fertilisation or treatments. However, local communities have lost the specific know-how to cultivate ancient wheat varieties. In addition, reproducing these varieties is often costly due to lower yield, particularly for small farms.

Most of the interventions named in Chap. 1 were needed for the success of this project, mainly incentives and capacity-building and cross-sectoral cooperation, with the most important leverage points being: visions of a good life; values and action; justice and inclusion in conservation; and education, knowledge generation and sharing.

Facing these challenges, this chapter discusses how a project run by a farmers' cooperative, the Grani Antichi Association (Ancient Grains Association), could bring about the seeds of transformative change based on the following: (1) maintaining a reasonable final cost for consumers; (2) raising consumers' awareness of the multiple benefits of local varieties; and (3) introduction of a sustainable producer-consumer cycle.

6.2 Background

6.2.1 Project Area: Montespertoli

The town of Montespertoli is located in the Florence metropolitan administrative area, in the region of Tuscany, at a distance of approximately 32 km south/south-west of the city (Table 6.1). The territory covers an area of approximately 120 km^2, and is crossed by the Elsa and Pesa rivers. The predominantly hilly terrain, made up almost entirely of sharp ridges and valleys with an average elevation between 200 m and 422 m, contributes to the fact that Montespertoli is less densely populated than neighbouring towns, which are located in flatter areas.

According to the 1871 census, there were 9135 inhabitants in Montespertoli, and today the population has increased to 13,500. Inhabitants are spread throughout the territory, an atypical situation compared to neighbouring towns where the population tends to gravitate to the town centre. The 1871 census reported that 38 families owned land in Montespertoli. Most of these families lived in Florence and sold farm products such as wheat, oil, wine and cattle. In 1871, 122 individuals served as public clerks, or in ecclesiastical or administrative positions, while the remainder of the population worked the land. Until the Italian economic boom of the 1960s, farming remained the primary activity in Montespertoli, with the addition of straw-hat production that employed approximately 1000 people, primarily women.

Over the last four decades, due to changes in external market conditions, industrial food production and mechanisation of farming techniques, a significant shift in land use has taken place, with a sharp increase in vineyards and wine production (22 km^2), a sharp decrease in land used for wheat production and the disappearance of cattle altogether. The higher cost of cultivating wheat in Montespertoli compared to wheat produced on flat terrain forced this transformation. This change also led to a shift in the social fabric, away from agricultural activities (including town fairs, harvest festivals, etc.), with a decrease in farmers and an increase in individuals who commute to the cities of Florence or Empoli for work.

Table 6.1 Basic information of the case study area

Country	Italy
Region	Toscana
District	Florence
Municipality	Montespertoli
Size of geographical area (hectare)	12,000
Number of indirect beneficiaries	13,474
Dominant ethnicity(ies), if appropriate	n.a.
Size of case study/project area (hectare)	12,000
Number of direct beneficiaries	n.a.
Dominant ethnicity in the project area	n.a.
Geographic coordinates (latitude, longitude)	43° 39′ 30.60″ N; 11° 3′ 32.36″ E

In Montespertoli there was a small SEPLS of wheat producers who sold their wheat to the only mill remaining in the area. In the past 50 years, all wheat production was converted from local grain varieties to just a couple of international varieties. However, as prices decreased, the SEPLS entered into a crisis as farmers could not earn sufficient funds to support their livelihoods. Under such circumstances, 6 years ago a project was introduced, aiming to reintroduce local varieties with higher revenues per kilogram to help farmers economically.

In this context, the gradual reintroduction of wheat production in Montespertoli starting in 2010 also contributed to the reintroduction of cattle and other animals in the area due to the rotation of wheat with foraging crops. At present, three cattle farmers are using 150 ha of pasture each, with one small local cheese production (*Marzolino di Lucardo cheese*) just beginning. As fields are used once again, more crops are being cultivated and more wild varieties can colonise the fields as no herbicide is used.

6.2.2   Project Activities

The key activities of this project include: (1) founding of an association, making it possible to incorporate cross-sectoral cooperation between universities, city councils, private companies, traders and farmers; (2) creating a new economic model with incentives and capacity building; and (3) replication and dissemination of the project with pre-emptive action. These activities were made possible by embracing a different vision of good life and ensuring environmentally friendly technology, innovation and investments through promotion of education and knowledge generation and sharing.

6.3   Founding of the Grani Antichi Association

6.3.1   Scope of the Association

As a response to local changes in wheat production in the Montespertoli area, the Grani Antichi Association project, which started in 2014, sponsors and verifies practices in relation to sustainable agriculture to increase the biodiversity of wheat, as well as other local flora and fauna. All members pay an annual fee to finance the activities and run the Association. The region and city council have also financed specific projects with ad hoc grants. The initiative is characterised by multi-sectoral cooperation among researchers from the University of Florence, members of the Montespertoli city council, millers, bakers, consumers, and farmers, who cultivate ancient and biodiverse varieties of wheat, such as Andriolo, Inallettabile, Sieve, Frassineto, Autonomia B, Verna, Gentil Rosso, Farro Monococco and Dicocco. These wheat varieties are grown, milled and processed locally. Some of these varieties would have disappeared without the activities of the Association, as they were no longer being cultivated. The Association is now the official custodian of several varieties. This means that they were replicated in members' fields, with a state inspector verifying that the replication process is done according to set rules. Each ancient variety was tested in a different environment and best practices were shared with farmers by the Association's technicians so that they did not plant the wrong type for their soil. The Association has also introduced other cultivation such as rye, hemp, chickpeas, fava beans and other traditional rotation crops. These ancient crops are taller, do not need fertilisers and are more resilient. However, they produce less and can be affected by some different pathogens that farmers need to be aware of (Fig. 6.1).

6.3.2 Patented Trademark

The Grani Antichi Association recognises and protects growers and processors (i.e. millers, bakers, and pasta makers) with a patented trademark. A specific logo serves as a guarantee that the bread, pasta and flour carrying that logo are made respecting the Association's guidelines. For this purpose, a Participatory Guarantee System (PGS) has been created. PGS is a locally-focused quality assurance system and formalises the adoption of the Association's guidelines for cultivation and processing of wheat products. It certifies wheat producers and processors based on the active participation of stakeholders (producers grow grains, processors mill and process them by cooking or other means, and consumers eat them) and is built on a foundation of trust, social networks and knowledge exchange. In the Association's case, an annual inspection of all members is conducted by a voluntary group of stakeholders (more on https://www.ifoam.bio/).

6.3.3 Wheat Product Analysis and Guarantee

The Grani Antichi Association, in cooperation with the University of Florence, randomly and selectively tests the wheat for any mould, toxins or pesticides. Any portion of the wheat that appears unhealthy due to the presence of diseases or anomalies is tested and discarded if necessary. A pathologist from the University of Florence also checks the wheat for pathogens in the field before harvesting. At any moment, consumers can visit the processors and verify their operations. In 2019, many small billboards were created to indicate to passersby which wheat fields were planted with ancient grains from the Association.

These measures increase trust between the Association and its producers. The notion of trust is understood as a belief in the integrity of the product and the process leading to the product, as well as the commitment of all stakeholders to the principles of the Association. A high degree of **acceptability** of the principles and philosophy

Fig. 6.1 Wheat field, Photo: Batistelli

of the Association amongst the various stakeholders is a fundamental factor in the project's success.

6.3.4   Grant and Funding for Farmers

If in the earlier phase of the Association, relations between the Association and stakeholders were based on trust, a more explicit and defined system of relationships has been developed in the current phase. Thanks to a grant of 619,273 EUR from the regional government (*Regione Toscana*) and a mutual 5-year commitment to the continuation of the project, ten members of the Grani Antichi Association were selected to contribute to developing the chain from producer to processor. The members have invested in machinery (specifically tractors) and in the establishment of a certified seed factory. One of the Association's members invested in the creation of a certified seed operation thanks to initial funding from this grant. Farmers can now provide seed samples, which will be cleaned with modern machines, stored and certified according to norms controlled by government specialists.

At the same time, the Association has invested in registering wheat varieties which do not currently appear in the national registry, in order to be able to sell them legally. Over the next 2 years, most varieties will be successfully registered. Many were cancelled from the national register when they were no longer traded. Once this registration process is complete and the seed company fully operational, the farmers will be able to buy and sell the ancient varieties with their specifically registered names, and no longer under the vague nomenclature of 'wheat'. This in turn means that the labelling system of the final products will report the specific wheat variety, increasing their commercial value and legally recognisable identity.

6.3.5   Education and Promotion

The Grani Antichi Association promotes the organic and sustainable cultivation of wheat based on the belief that it is better for the environment and healthier for consumers. Members of the Grani Antichi Association have focused their activities on speaking and writing about the benefits of ancient grains for consumers, farming best practices and the economic implications of the project model. Conferences, participation in food markets, lessons in schools, universities and town halls are frequently carried out. Activities have also included lobbying with members of the public sector.

In partnership with the University of Florence, conferences have been held at medical centres with the aim of disseminating results of the latest research on ancient grains, which illustrates the health benefits of using ancient grains over commonly-used modern wheat varieties (e.g. Sofi et al. 2010). A recent conference at the University of Florence had the positive result of convincing one agronomist in Southern Tuscany that starting a project in his area to convert intensive agricultural operations to sustainable cultivation of ancient grains was viable. After learning about the ancient grains project, this individual then lobbied his own local city council to sponsor a project in his area.

The Grani Antichi Association has also organised lectures and events in different city councils, mostly at the invitation of council members. Councils looking for solutions to improve local agricultural problems, convert abandoned areas, or improve local agricultural production have contacted the Association with invitations to illustrate that cultivating ancient grain wheat varieties in a sustainable and organic way is possible and economically viable. Resistance to the idea exists, especially after decades of conventional agriculture, so the Association uses data and testimonials to demonstrate the viability of projects of this sort on many levels.

6.3.6 Challenges

The survival of the Grani Antichi Association depends on the trust that has been created among its members, and between members and the local community. Most farmers tend to distrust food processors and large mechanised farmers. That is why it is important for all stakeholders to share a consensus on the principles and objectives of the Association. So far, the miller involved in the Montespertoli project has experienced an upswing in business. The environment prior to the project was challenging due to large mills gaining a greater share of business at lower prices. More than half of the miller's business now involves ancient grains. The three bakers involved have also increased their business by 10%, 20% and 100% respectively in three years. Similarly, the pasta makers and other farmers who sell ancient grains directly have increased their business and created new opportunities. One former member of the Association created a new agro-company which deals only with ancient grains, replicating the whole chain of production. Many people have been hired as a result of the project, and several businesses that had not been doing well are now thriving. During the COVID-19 lockdown, most of the farmers and processors selling ancient grain products sold the same amount or more products (some as much as double). Overall, they have become more resilient. Ancient grain clients generally buy regularly. The city council and many food processors are also using the ancient grain logo in their communication to gain prestige as, locally, it has been recognised as an added value. The Association is currently discussing the extent to which members are allowed use the logo in promoting their activities. Results have not yet been disclosed.

One of the main challenges the Association has faced has been how to keep the operation to a manageable size. In 2019, the Association defined and maintained a maximum of 10 wheat-producing hectares per farmer member per year. This allowed for the total number of members to increase, without increasing the total wheat production of the area. Without a maximum, a small number of large landowners could have produced a significant percentage of the overall wheat production, excluding newer, smaller farmer members from the Association. However, the Association had to manage the discontent of those farmers with larger *potential* production maximums. This contrasts with other models where the tendency is for a small number of large producers to emerge as the main stakeholders and decision-makers. *The priority of community participation was valued over individual margins for growth.* In addition, the Association has had to take into account the limited number of processors and limits to the client base in balancing the demand for wheat and the size of the production chain. Accordingly, the Association decided not to increase the total production of wheat volume per year, but rather to support replication of the project, both in proximity and in locations further afield. Some of the conditions for, and considerations involved in, the process of scale-up are considered below.

Another challenge concerns the duties and rules of food processors. Several documents and guidelines are shared by food processors who prefer loose rules and business friendly guidelines. Their product lines made with ancient grains are limited due to rigid protocols. So far, the Association has refused to open up to bigger companies that would increase the demand for ancients, but would also want more freedom. This is a limitation to the growth of ancient grain cultivation within the Montespertoli project, and it is not likely to change soon.

6.4   Creating a New Economic Model

One of the most important issues, if not the most important, for the successful implementation of a model, such as the one used by the Grani Antichi Association, is the economic implication. Looking at the activity of the Grani Antichi Association, it is clear that the market economy model does not fit a socio-economically sustainable grain system. Farmers have fixed costs and high risks, and fluctuating prices rarely correspond to these costs. In most of the systems involving ancient grains in Tuscany, prices of products along the supply chain are set either by the initiator of the project or with the help of the University of Florence. Five of these models, including the Montespertoli ancient grains, have already been studied and analysed (Sacchi et al. 2019). In this section, we will illustrate the Montespertoli case.

6.4.1   The Market Economy Model

While demand and supply have traditionally influenced wheat prices, climatic conditions and crude oil prices are now two important factors affecting wheat production. As stated by Enghiad et al. (2017), "Oil prices influence the cost of inputs for wheat production, and similar patterns observed in wheat and oil price fluctuations indicate high correlation between the two". Montespertoli farmers believe that a market economy model, where global supply and demand determine price, is not a viable model in many areas of Italy, where costs are higher than in other nations due to historical and geographical specificities (i.e. small, separate farms lots, and hilly terrain).

6.4.2 Possible Models

Apart from the standard market economy model, where farmers try to sell to anybody they can for the best possible price at a given time, several other models exist. In Tuscany, some farmers belong to a consortium, where the consortium determines annual wheat prices and annual quantities for a group of farmers based on the consortium's ability to sell the wheat to distribution systems. Prices are set according to market demand, but farmers have some level of protection as at least annual prices and quantities are respected. Some other models involve a company or farm which subcontracts production to other farms. Figure 6.2 shows the different models that were studied by Sacchi et al. (2019).

Normally, a farmer sells wheat to a miller who then transforms it into flour and sells it to bakers. Bread is usually sold by bakers to shops or directly to consumers. In Tuscany, supermarkets sell mainly industrial bread and local bakeries sell mainly artisanal bread near residential areas. The graphic in Fig. 6.2 shows the share at different stages of the premium of bread, in EUR per kg. Montespertoli redistributes benefits more equally and maintains a reasonable pricing. The other chains are different Tuscan producer-consumer chains from the more industrial (lower price) to artisanal (high price). Sacchi et al. (2019, pp. 5–6) report,

> While B1, B3, and B4 redistribute a relevant part of the premium over conventional bread to farmers, B2 and B5 mainly favour the final stages of the production chain. In B2, the gap between the purchase price of the flour by bakers and the final selling price of the bread is much wider, allowing bakers and retailers, such as large-scale distribution networks, to obtain a higher gross margin and quota of participation in the premium. The B5 chain, in

Fig. 6.2 Benefit distribution of bread revenue in different chains in Tuscany in EUR per kg (Sacchi et al. 2019). *Note.* Figures for Montespertoli are shown in Chain B3. The other chains are industrial, supermarket and artisanal

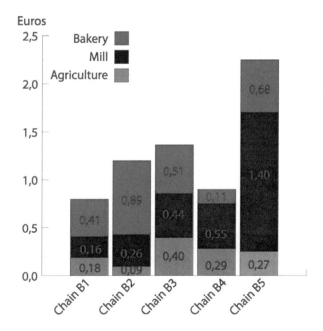

Table 6.2 Targets and indicators for the Montespertoli SEPL

Indicator	2013	2019
Montespertoli certified grains harvested	1000 kg	300,000 kg
Ancient grains harvested	n/a	250,000 kg
Wheat varieties cultivated	4	28
Farmers involved	3	45
Ha cultivated	10	500

Table 6.3 Amount of ancient grains farmed and milled and number of wheat varieties increased in Italy

Indicator	2013	2019
Ancient grains harvested	n/a	500,000 kg
Wheat varieties cultivated	5	50
Farmers involved	10	100
Hectares cultivated	200	1,000
Communes involved	2	15

contrast, favours the milling stage in particular. As far as the distribution of the price premium is concerned, B3 is the chain with the most equitable distribution (...).

An example of a very small-scale model is a farmer that cultivates the product, transforms the product and sells it directly to consumers. Even though this is possible, it requires an array of different types of knowledge, the right range of equipment and the involvement of several people. Not many farmers can achieve this model. In Montespertoli, a few farmers (3) have chosen to work this way, using the Association brand name. They are producing, processing and selling in local markets to a small and faithful consumer base, on their own.

In other cases, sometimes overlapping with the above-mentioned examples, community farming associations or consumer associations have managed to skip middlemen or shops. In those cases, there has been an increase in farmers' average selling price (ASP). However, these systems seem to have plateaued and usually represent a small niche in consumer buying patterns. In the area of Montespertoli, individuals generally purchase their bread and pasta at a shop or local supermarket, which requires products to be delivered to these locations.

Table 6.2 shows partial indicators for ancient grains SEPLs in Montespertoli. In Italy, there are many more hectares today (more than 10,000), but data is not available as ancient varieties were not accounted for as they were not in the national registry in 2013.

Table 6.3 shows partial indicators for ancient grains SEPLs in Italy, from projects directly sponsored by the Grani Antichi association from Montespertoli.

6.4.3   Fixed Price of Wheat

Addressing the question of how to guarantee that farmers' costs are covered while also maintaining a market price that is attractive for shops and consumers, can be considered the Grani Antichi Association's most innovative achievement. The Grani Antichi Association first established a minimum revenue per hectare that would allow farmers to cover costs and earn a small amount. From this cost base, the price of bread was determined so that both millers and bakers would make a profit. To ensure that the product was within the purchase range of many people, rather than merely a wealthy elite, the final part of the chain is not allowed to raise the price more than a certain amount to guarantee the affordability of the product for most people. In this way, a fair distribution of profit was reached with approximately a third of the total price given to the producer, a third to the processor and a third to the shop (Fig. 6.2).

6.5   Replication in Other Settings

6.5.1   Implementation Research

Implementation research looks at how to strengthen systems using the experiences of stakeholders (decision-makers and implementers), while taking into account local context, priorities, and the complexity of the systems. Implementation research addresses an array of issues, such as acceptability, adoption, appropriateness, feasibility, fidelity, implementation cost, and sustainability, and is meant to address the 'know-do' gap.

In the case of the Grani Antichi Association project, and subsequent replications of similar projects in different Italian contexts, initial research into factors that have affected the success of the initial project (above), and have emerged as significant in other contexts at the outset is presented. Given the very recent developments beyond the first project, observations are of a preliminary nature, and require further research as each project grows over time. The Association has played an active role in encouraging farmers to replicate a similar system in other settings as an alternative to industrial wheat production.

6.5.2   Initial Observations of a Modest Scale-Up

We have considered the categories of acceptability, adoption and sustainability in our preliminary analysis. Acceptability refers to the community's belief in the principles of production and consumption outlined above. This includes various stakeholders: farmers, millers, processors and consumers. Adoption includes adherence to the growing guidelines (i.e. crop rotation, elimination of fertilisers, public information, control visits, etc.). Finally, sustainability includes factors that relate to the viability of the project over time from the perspective of each stakeholder. Qualitative data was collected through interviews conducted during the period from February to November 2019.

6.5.2.1 Implementation in the Umbria Region

A local Slow Food group learned of the Montespertoli experience and acted as a facilitator in creating a new group in the Umbria region. The group visited the Association in Montespertoli and obtained seeds and initial know-how to grow them and mill the wheat. In terms of acceptability, because the Slow Food group members already shared the Slow Food principles of 'good, clean and fair', acceptance of ancient grains and the economic model of 30% for each part of the chain did not meet with resistance. This initial adoption, even if it is small in volume in terms of surface area converted and grain output, could be significant in terms of visibility as it is implemented by a Slow Food group. If successful, it could serve as example for other farmers in the area.

In terms of adoption, a single farmer adhered to ancient grain growing methods, and supplied all of the wheat for one miller, who purchased it at a pre-set price, and processed the wheat for the Slow Food community. The market demand was defined by the pre-existing community, and the farmer and miller adhered to that quantity. A preliminary analysis views sustainability as relying on a continued and steady demand that does not exceed the farmer's land use and annual production. (size: 2 farmers, 10 ha)

6.5.2.2 Implementation in Pavia

In Pavia, two farmers who had lived in Montespertoli acted as a lever for the farming community there. In the hills of Pavia (Oltrepo Pavese), unlike other nearby areas with prestigious wine denominations which lead to growth in vineyard cultivation, no lucrative agricultural activity existed. It was therefore easier to convince a group of farmers to convert to ancient grains cultivation.

Adoption of the project included seven farmers, two millers and a number of bakers who undertook a radical change in their practice. Unlike the Umbria group, the Pavia group did not initially include consumers. Market demand needed to be created through local community education.

Therefore, in terms of acceptability, the initial farmers and processors involved believed in the viability of the project based on the principles of improving the quality of the wheat, increasing biodiversity and introducing sustainable practices. They did not, however, have a guarantee that the community would show the same level of acceptance. Sustainability would depend on the consumer demand that supports the production on the seven farms.

After their first year of production, the entire yield was not sold. However, another miller from Val di Susa, in the Piedmont region, purchased the excess wheat and intends to begin his own project in his local area. (12 farmers, 20 ha)

6.5.2.3 Implementation in Castelfranco di Sotto (Pisa Area)

The mayor of Castelfranco invited the authors to hold a conference with the farming community in an aim to introduce sustainable practices and convert an economically depressed, conventional agricultural area into an organic farming district with biodiverse and healthier crops.

A large part of the land was owned by four farmers who were initially skeptical about the guidelines for cultivation of ancient grains, as well as the viability of the entire project which is based on selling wheat at higher prices than conventional wheat. The mayor and the vice-president of a local food association (Centro Commerciale Naturale) decided to focus specifically on the farmer with the largest farm, who was also considered a trend-setter in the community. After attending the conference, meeting with members of the Association and learning about how the project was implemented, the farmer was willing to try. The other farmers will wait to see the results of the trial. According to the local association, if the lead farmer decides that the project is viable, the others will follow.

6.5.2.4 Implementation in Sansepolcro

After a workshop about sustainable and organic agriculture (AIDA: *ASSOCIAZIONE ITALIANA DI AGROECOLOGIA*, or Agro-ecology for Organic Agriculture, University of Florence, 15 November 2019) held by the authors, an agriculture consultant decided to propose a workshop in Sansepolcro with local farmers. The mayor hosted and encouraged farmers to participate. A meeting followed, and several farmers were convinced by Montespertoli members. These farmers had been mainly cultivating tobacco as a cash crop and had very low margins. They decided to make a change and introduce more sustainable and organic practices which would benefit the local community; however, given that no bread or flour chain existed, the farmers decided to begin by producing wheat for pasta only.

Similar to the Pavia example, in terms of acceptability, the farmers involved believed in the viability of the project based on the principles of improving the quality of the wheat, increasing biodiversity and introducing sustainable practices. Sustainability will depend on consumer demand and acceptance of the principles of the project (several farmers involved).

6.6   Discussion: Transformative Change Starting from the Grain Farmer's Association

The redistribution of economic benefits in the context of the Grani Antichi Association's activities has altered the production and transformation chain with beneficial effects on the ecosystem, as well as a redistribution of money. This has created positive outcomes as well as tensions in the group benefitting from the change. After the work of the founding members, who were primarily motivated by commitment to sustainability and biodiversity, farmers in Montespertoli and in the other areas took up the practice mainly because there is a fair distribution of the economic benefits (Fig. 6.2). It is clear that economic motivation is key in most cases for farmers. For politicians, however, support and consensus are very important.

The Association continued to grow in number of farmers until the mill reached full capacity. Now the production has hit a plateau and aspiring members have been encouraged to start an association of their own, which has happened in many cases (see below and Table 6.3). As new associations are being created, indicators for Italy are growing, while Montespertoli's remain stable.

6.6.1   Consumers As Co-producers

Normally, supermarkets have purchasing managers who search for products at the lowest prices. Producers are therefore pushed to have low prices and try to adapt by selecting providers of agricultural products at the lowest price in turn. For other products, especially niche ones, rarity or brand name can permit the reseller to have higher prices. In terms of wheat, ancient grain wheat and organic wheat are seen as more valuable as they constitute a healthier and more sustainable option, and are therefore sold at higher prices. However, that high price is not always transferred back to the producer. As the ancient grain chain in Montespertoli fixes prices for processors and resellers, the benefits are equally distributed (one third respectively to producers, processors and resellers). In this scenario, the end users (the consumers) are involved in checking the chain via the PGS and feel they are part of the chain as they are informed about the relationship between production and costs.

6.6.2   A Leap of Faith

When consumers are considered part of the chain, they need to trust that the organisation and the farmers are not trying to take advantage of them through pricing. As mentioned above, building trust is an important aspect of the Association's experience. The action of identifying each field with a metal sign with the name of producer, type of wheat planted and rotation crops is one example of this.

Regularly planned community meals occur where people can participate and meet the producers. Every year a harvest celebration is another occasion to meet the producers and taste the products. Meetings and assemblies are always open to the public and advertised locally.

6.6.3   Lessons Learned in Montespertoli

One significant lesson learned in the past 2 years is that overproduction is more destabilising than wheat shortages. If there is a shortage, demand for products increases. Farmers sell the entirety of their product, and though they might not make the whole amount projected because of the shortage of wheat, they accept it. On the contrary, if farmers produce more wheat than expected, they saturate the market, but the price for wheat cannot decrease because it is set. The processors cannot therefore purchase all the wheat harvested in that given year, and projected purchasing agreements are compromised.

In 2019, the wheat yield was 50% more than projected. Accordingly, the Association improved its quota system by putting a limit on both the hectare and the total number of kilograms per producer. Previously, the quota was based on hectares only. The quota system was proposed by the board of directors and voted on by members. In 2020, after a suggestion from the board, the farmers decided to assign the same quota to everyone regardless of any factor, such as small or large farms, founding or most recent members: they all cultivate three hectares and sell at the set high price of 4.5 tons of wheat. After a process of discussions, this emerged as the fairest choice. It was possibly the biggest adaptation of the project. It remains possible to exchange this right to sell between farmers at their convenience. This egalitarian choice has not been replicated in other projects so far. Overproduction in other areas has been solved primarily by lowering prices, similar to market economy mechanisms.

After many years of trials, farmers also learned that production can be abundant even without fertilisers or chemical products. Converting abandoned fields or industrially farmed fields to ancient varieties has improved water management (by digging ditches), landslide control (by planning crops that contain erosion), and production sustainability (fertile soil without chemicals).

6.6.4   Scale-Up of the Project

Acceptability of the principles of the Grand Antichi project must reach beyond the political or social, and include the producers, if the project is to take hold. This is logical, as the farmers must dedicate a portion of their land and resources. The question of consumer acceptability is variable, as some projects have started with a strong consumer base, and others have yet to create one, but are confident that the experience in Montespertoli can be extended to their areas in terms of consumer interest, buy-in and support.

Adoption can be quite limited, as the case of Umbria illustrates. However, adoption is linked to sustainability in each case, as the number of farmers and the amount of grain produced must be absorbed by the local market if the project is to maintain its commitment to a 30% partition of earnings.

A significant factor in the replication process and the level of success of a project appears to be the level of commitment of the main stakeholder as project driver. Adoption of clear policies is also seen to contribute to success to a lesser degree. The education and motivation of the farmers are fundamental factors as well. Last but not least is the creation of an economic model that includes consumers. Consumers need to be educated that by purchasing products made locally in a sustainable way they are sponsoring cleaner agriculture that improves the soil and redistributes revenues in a fairer, healthier way, which is the base of transformational change.

Based on the results and experiences of this project, the following quantitative indicators that measure the progress of these activities in the aim for transformative change are suggested:

- Total number of projects in biodiverse grains in Italy;
- Number of direct farmers involved in different communes in Italy year to year;
- Number of hectares cultivated through local association activity;
- Number of kilograms of wheat harvested;
- Average price per kilogram of wheat paid to farmers;
- Number of government incentives for cultivating biodiverse wheat;
- Number of city council policies favouring organic agriculture of biodiverse wheat; and
- Number of varieties in national system reporting and registry.

Some of these values are reported in Tables 6.2 and 6.3 in relation to the Montespertoli project, 2013 to 2019. It is expected to provide a comparative base for future analysis and support for not only further scale-out but also scale-up and scale-deep.

6.7 Final Reflection: Changes in the Community

Normally people shop looking for what they like at the lowest price. However, the new system described herein requires people to understand that low prices have repercussions up the chain, and local people cannot bear the burden of price instability. Consumers need to step out of market economy rules and adapt to paying prices that reflect the actual cost of the product, knowing that the benefits are equally redistributed along the chain. In this way, consumers become project participants rather than just end users. This involves concerted local communication and social efforts to involve consumers so they see themselves in this new role. It also profoundly changes individuals' feelings of responsibility leading to positive action regarding sustainable practices and concrete responses to climate change.

Another significant change was the realisation within the Association that making cultivation of ancient grain wheat attractive through rewarding producers can lead to expectations that these new benefits are guaranteed to the individual in all circumstances. If, as in the case of a limited distribution chain and excess yield, even beyond the Association's mitigation plan of surface and plant allocation, adjustments need to be made, the project risks collapse because of negative feelings among some producers. In 2019 in Montespertoli, benefits could only apply to a few producers with large cultivated surfaces **or** to many producers, including those with small surfaces. The Montespertoli Association decided to benefit a large number of producers each with small areas, negatively impacting the larger producers who could potentially have sold more.

Replication requires an understanding of the relationship between production and demand, and also consideration of who the main stakeholder is during the implementation of the project, as different stakeholders with different roles will introduce different limitations to the project. For example, in Montespertoli, the first and strongest stakeholder was the miller, and the chain was primarily organised around the quantity of wheat that the mill was able to transform. Recognising this limit, a second mill was introduced thanks to an investment made possible by a grant from the Tuscan region. In some other chains, for example where the initial and strongest stakeholder is a farm using mills as an external resource, the problem is primarily market demand. If, on the other hand, the main stakeholder is a shop, the chain will be created based on its market potential.

To conclude, projects of this kind require consumers to change their purchasing attitudes. They need to become co-producers as project participants and understand the cost and value of agricultural products. In this way, superior products are made available and a socio-ecologically sustainable environment is maintained. The producers as well as the processors need to create a common chain so that they all benefit from consumer awareness.

Ultimately, the biggest goal would be to transform society into groups of people who understand that humans have a responsibility to preserve biodiversity and the environment. Therefore, all fields would be cultivated in a sustainable way, and all bread would be made with wheat made in those fields, as people would not want any other. This goal applies to all other agricultural products as well. The Ancient Grains Association's actions are a concrete response to Goal 12 of the Sustainable Development Goals by the United Nations on *Responsible Consumption and Production* (Fig. 6.3).

Fig. 6.3 Harvest party (Photo: Batistelli)

References

Enghiad, A., Ufer, D., Countryman, A., & Thilmany, D. (2017). An overview of global wheat market fundamentals in an era of climate concerns. *International Journal of Agronomy, 2017,* 1–15.

Sacchi, G., Belletti, G., Biancalani, M., Lombari, G. V., & Stefani, G. (2019). The valorisation of wheat production through locally-based bread chains: Experiences from Tuscany. *Journal of Rural Studies, 71,* 23–35.

Sofi, F., Ghiselli, L., Cesari, F., Gori, A. M., Mannini, L., Casini, A., Vazzana, C., Vecchio, V., Gensini, G. F., Abbate, R., & Benedettelli, S. (2010). Effects of short-term consumption of bread obtained by an old Italian grain variety on lipid, inflammatory and hemorheological variables: an intervention study. *Journal of Medicinal Food, 13,* 1–6.

Chapter 7
Sustainable Rural Development and Water Resources Management on a Hilly Landscape: A Case Study of Gonglaoping Community, Taichung, ROC (Chinese Taipei)

Chen-Fa Wu, Chen Yang Lee, Chen-Chuan Huang, Hao-Yun Chuang, Chih-Cheng Weng, Ming Cheng Chen, Choa-Hung Chang, Szu-Hung Chen, Yi-Ting Zhang, and Kuan Chuan Lu

Abstract The Gonglaoping community is located in Central Western Taiwan, with approximately 700 residents. The hilly landscape contains farmlands and sloping areas with abundant natural resources. Locals rely on the Han River system and seasonal rainfall for water supply for domestic use and irrigation. Uneven rainfall patterns and high demand for water has led to the overuse of groundwater and conflicts among the people. The surrounding natural forests provide important ecosystem services, including wildlife habitats and water conservation, among others; however, overlap with human activities has brought threats to biodiversity conservation. Considering these challenges, locals were determined to transform their community towards sustainability. The Gonglaoping Industrial Development Association (GIDA) and the Soil and Water Conservation Bureau (SWCB) joined hands to initiate the promotion of the Satoyama Initiative, playing catalytic roles in several implementations, such as establishing water management strategies based on mutual trust, rebuilding the masonry landscape, and economic development,

C.-F. Wu · Y.-T. Zhang
Department of Horticulture of National Chung Hsing University (NCHU), Taichung City, Chinese Taipei

C. Y. Lee · C.-C. Huang · H.-Y. Chuang · M. C. Chen
Soil and Water Conservation Bureau (SWCB), Nantou City, Chinese Taipei

C.-C. Weng · C.-H. Chang
SWCB Taichung Branch, Taichung City, Chinese Taipei

S.-H. Chen (✉)
International Master Program of Agriculture, NCHU, Taichung City, Chinese Taipei
e-mail: vickey@dragon.nchu.edu.tw

K. C. Lu
Gonglaoping Industrial Development Association (GIDA), Taichung City, Chinese Taipei

© The Author(s) 2021   115
M. Nishi et al. (eds.), *Fostering Transformative Change for Sustainability in the Context of Socio-Ecological Production Landscapes and Seascapes (SEPLS)*,
https://doi.org/10.1007/978-981-33-6761-6_7

forming partnerships with other stakeholders. This multi-stakeholder and co-management platform allowed the community to achieve transformative change, particularly in resolving conflicts of water use, restoring the SEPL, enhancing biodiversity conservation, and developing a self-sustaining economy.

Achieving sustainability in a SEPL requires the application of a holistic approach and a multi-sector collaborating (community-government-university) platform. This case demonstrates a practical, effective framework for government authorities, policymakers and other stakeholders in terms of maintaining the integrity of ecosystems. With the final outcome of promoting a vision of co-prosperity, it is a solid example showing a win-win strategy for both the human population and the farmland ecosystem in a hilly landscape.

Keywords Agricultural landscape · Eco-friendly farming · Water resources conservation · Dry stone masonry · Sustainable rural development · SEPLS

7.1 Introduction

Gonglaoping, a hillside rural community in Fengyuan District, is situated in the central western part of Taiwan Island (Fig. 7.1), and has approximately 700 residents (Table 7.1). The main landscape is a terraced terrain with a total area of about 250 ha, containing approximately 100 ha of farmlands and 80 ha of foothills with abundant natural resources. The entire community is located within the upstream watershed of the Han River. River terraces are mainly formed by the Toukoshan Formation that originated from an alluvial plain and graduated to its current status due to orogenesis. The topography contains some flat areas with a mean elevation of 450 m. Thus, most residents have settled in flat areas but cultivate orchards on sloping hills. The geological origin of soil on the hills is from the Cholam Formation and lateritic river sediments. The Cholam Formation contains sandstone, mudstone and shale while river sediments consist of red clay, gravel, sand and other sediments. Both formations present good permeability.

The mean annual temperature of Gonglaoping is 21.6 °C, and weather patterns show distinct seasons with uneven rainfall patterns. The wet season starts in April and lasts until September with a mean monthly precipitation of 268 mm, while the dry season runs from October to March with a mean monthly precipitation of 42 mm. In recent years, droughts have become more frequent due to the impacts of climate change. Consequently, the terrace topography, low water retention capacity in the soil and unpredictable weather patterns have led to local production landscapes becoming increasingly vulnerable to the effects of regular droughts.

Due to uneven patterns of precipitation (e.g. the dry season occurring from October to March), the amount of water is insufficient to fulfil local demands; as a result, farmers extract groundwater for irrigation. Alongside climate change impacts, droughts are getting more and more serious as the dry season is lasting longer and leading to over-pumping of groundwater upstream. This situation threatens to

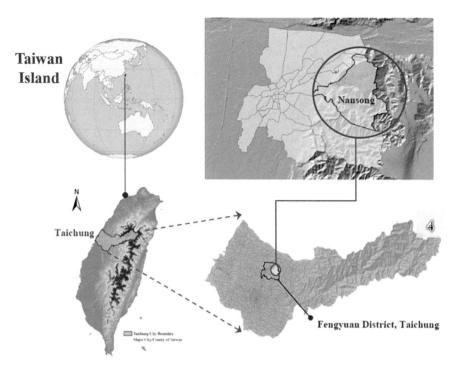

Fig. 7.1 Map of the country and case study area. Counter-clockwise from top left, relative location of the island of Taiwan (from Wikipedia), Taichung City (rose-coloured area), Fengyuan District (red outlined area), and Nansong Village (brown outlined area), where the Gonglaoping community (light orange-shaded area) is situated

Table 7.1 Basic information of the study area

Country	Chinese Taipei
Province	Taichung City
District	Fengyuan District
Municipality	N/A
Size of geographic area (hectare)	4,118.45
Number of indirect beneficiaries	166,688
Dominant ethnicities (if appropriate)	Han Chinese
Size of case study area (hectare)	250
Number of direct beneficiaries	685 (2018 census)
Dominant ethnicities of the case study area (if appropriate)	Han Chinese
Geographic coordinate (Latitude, Longitude)	24° 15′ 33.12″ N; 120° 45′ 53.64″ E

jeopardise the water rights of humans and other fauna species living in midstream and downstream areas, and moreover causes water-related conflicts among community people (SWCB Taichung Brach, 2017).

Fig. 7.2 Bird's eye view of persimmon orchard landscape (Source: NCHU)

About 200 years ago, Shiu Gong arrived to the area as a pioneer to explore the land and start cultivation. Afterwards, settlements began to form on the plain near the headstream of the Han River. The land near the headstream received sufficient water supply; however, the terrain's slope, ranging from 32° to 52°, made it hard to further expand agricultural activities. To stabilise the hillside soil, the ancestors collected available materials (e.g. stone, and pebbles) in or near the area for construction. Stones were piled up in the form of retaining walls and formed into dry masonry embankments in the valley, which over time formed an iconic socio-ecological production landscape (SEPL).

Prior to the 1940s, locals mainly grew food crops, such as sugar cane, rice and potatoes for subsistence use. After 1945, Chinese Taipei's GDP started to increase and the economy developed quickly, causing market demand for food crops to plummet. Given that the micro-climate in Gonglaoping is quite suitable for the growth of fruit trees, farmers began to convert farms to orchards, with trees such as citrus fruits, persimmon and lychee. Persimmon, a temperate fruit tree, is typically grown on the hilly farmlands. Persimmon orchards reveal different colour themes by seasons. In spring, a cluster of green leaves grows on the thin branches, while in summer, trees start fruiting. In autumn, the persimmons gradually mature, and the fruit itself turns dark orange. When winter comes, the hibernating period of the persimmon trees begins, with the colour of leaves turning from green to red or golden yellow. The whole hillside shows a unique production landscape mosaic with masonry stone embankments and vibrant red-golden-yellow colour (Fig. 7.2) (SWCB Taichung Branch, 2017).

In 1990, farmers in Fengyuan District established the 5th Citrus Agricultural Production and Marketing Group. The group often holds training workshops or forums for members to exchange experiences and learn from each other. In order to reduce production costs, members set up a joint venture agreement to purchase a citrus fruit cleaning and sorting machine, refractometers (Brix meters) and other instruments. They also invited online marketing experts to teach members how to operate e-commerce. In this way, farmers gained the capacity to allow them to engage in direct sales online, reduce the cost of agency sales, access consumers' demands and feedback effectively, and increase profit margins. In 2009, the 5th Citrus Agricultural Production and Marketing Group of Fengyuan District won the national top-ten prize awarded by the Council of Agriculture, Chinese Taipei (COA). Owing to the high-quality fruit and brand name recognition, the sale values and prices of the fruits increased substantially, benefiting local farmers directly through increase in income.

Urban dwellers often have more job opportunities and better salaries. Since the Gonglaoping community is not far from downtown Taichung, young people tend to migrate to the city, causing problems such as an aging labour force and industrial recession. Currently, residents of Gonglaoping aged 60 or above account for 20% of the total population. In other words, the community has fallen into the pitfall of an aging labor force due to youth migrating out.

To respond to the challenges mentioned above, the Gonglaoping community committed to seek foster and apply a plan that could transform the SEPL toward a sustainable future. Locals were also determined to preserve and pass down traditional culture that emphasises humans living in harmony with nature. The embedded culture and beliefs have not only strengthened the sense of coherence but also have aided the community in establishing a system based on mutual trust, which has been beneficial to achieving the desired results. Therefore, it was quite essential to have a holistic approach that could sustain equilibrium among agricultural activities, water resources management, biodiversity conservation, local economic development and quality of life, as well as an approach that operated based on co-management.

7.2 Description of Activities

Transformative change refers to ". . .*fundamental, system-wide change that includes consideration of technological, economic and social factors, economic, and social factors, including in terms of paradigms, goals and values*" (Bélair et al. 2010; IPBES 2019). Referring to the principles above, the Gonglaoping community, National Chung Hsing University (NCHU) and the Soil and Water Conservation Bureau (SWCB), a government agency, joined hands to implement a project based on the Satoyama Initiative, to enable transformative change towards the desired goals of wellbeing and sustainability through several leverage points, which are:

(1) establishing water management strategies and a co-management system;
(2) restoring the masonry production landscape (SEPL);
(3) enhancing biodiversity conservation;
(4) developing products/services for a sustainable economy; and
(5) forming a multi-stakeholder operating platform.

Under such a multi-sectoral, collaborative structure (government-university-community), an array of actions has been carried out to achieve sustainability of the SEPL. These include forest and wildlife-habitat restoration in the upstream catchment area, eco-friendly farming practices to maintain healthy agricultural ecosystems, promotion of smart allocation and mitigation of over-use of water resources, preservation and re-introduction of traditional wisdom (e.g. dry stone masonry), and development of community industry to enhance local incomes. Related activities are described in the following sections (SWCB Taichung 2017; SWCB 2018).

7.2.1 Establishment of Water Management Strategies and System

Due to the unbalanced nature of seasonal precipitation, the amount of water available is often insufficient to fulfil local demand in winter. As a result, over-pumping of groundwater occurs, causing water-related conflicts among community people. Accordingly, the community requested funding and technical support from SWCB. SWCB took the lead role (catalyst) in creating a plan to dredge severely silted ponds and build up reservoirs. The total volume of reservoirs was expanded from 56 to 416 tons. The reservoirs store rainfall in the wet season to supply irrigation water in summer and winter. As shown in Fig. 7.3, the smart water resources management and recycling system allocates water resources precisely and effectively. For example, if the water level of the first and second ponds drop, the system would flag a signal to the pumping motor and the water in the downstream number three pond would be pumped up and distributed to the first and second ponds. Additionally, the pumping system is powered by solar energy, a clean energy that does not produce any greenhouse gases, known to be responsible for climate change.

To mitigate low soil water retention capacity, vegetated buffer strip cultivation was adopted to improve the physical and chemical characteristics of soil, in particular, to increase organic matter content thereby improving soil drainage and aeration. This approach can also prevent root damage due to sudden changes in soil temperature. Grass roots also retain soil moisture and nutrients, thus preventing rapid leaching of nutrients. The decomposition of grass roots supplies a large amount of carbon and nitrogen to beneficial micro-organisms, which enhances the soil's microbial cycles. In Gonglaoping, vegetated buffer strip cultivation is practices on 91% of farmlands. The other 9% is sloped land that can be difficult to artificially weed. However, the orchard owners counted on the advantages of vegetated buffer strip

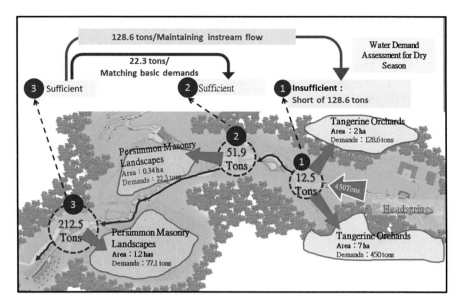

Fig. 7.3 The water resources management and recycling system for the Han River's upstream catchment area

Fig. 7.4 Formation ceremony of the Han-River Neighbourhood Watch (Source: GIDA)

cultivation to control weeds without using herbicides. Farmers have recognised that vegetated buffer strip cultivation improves soil properties, and contributes to environmental and ecosystem protection.

The community formed the Han-River Neighbourhood Watch that patrols the area regularly to voluntarily guard the headspring water, monitor the streamflow, and to coordinate water resources allocation (Fig. 7.4). Major duties include, but are not limited to: (1) monitoring of ecosystem health; (2) maintenance of a clean community; and (3) accounting for the regular cleaning and maintenance of water reservoirs. The neighbourhood watch is also responsible for the maintenance of weather stations and water level gauges, as well as data collection and monitoring. Community members also attend regular meetings, report maintenance results, exchange information on experience, and deal with any water-related issues on a weekly basis (every Tuesday night). Through this process, locals were able to foster a co-management system based on mutual trust to resolve conflicts and use water effectively.

7.2.2 Restoration of Masonry Production Landscapes

On 21 September 1999, a magnitude 7.3 earthquake occurred in central Taiwan, becoming the most devastating natural disaster in the island of Taiwan in the twentieth century. The Gonglaoping community was traumatised. Homes were damaged; masonry embankments collapsed, and roads and hiking routes were destroyed. After the earthquake, the community initiated a series of post-disaster recovery and reconstruction projects. In order to strengthen community cohesion and set clear development goals, the Gonglaoping Industrial Development Association (GIDA) took the lead in the implementation of a Rural Regeneration Project funded by SWCB. Likewise, SWCB initiated the Rural Empowerment Project in 2004 aimed at enhancement and capacity building for rural communities. Under this project, rural communities were requested to analyse their requirements and demands first, then determine their implementation capacity and make a plan for training to narrow the capacity gap. In April of 2010, the Gonglaoping community began the procedure to plan training courses. Subsequently, a four-stage, 92-h rural community development course was implemented for community members. In particular, the course taught how to document community life, agricultural production, ecology, culture and other resources, and then facilitated the development of a vision and goals independently. Through this capacity building programme, the community was empowered and the benefits of a participatory, community-led approach were clearly understood. In late 2012, the community completed and got approval for its rural regeneration plan proposal. After implementation of the rural regeneration plan, the Gonglaoping community actively and independently applied government funds to improve the community environment, conduct cultural and ecological surveys, implement community guide training programmes, and organise activities to invigorate the citrus fruit industry every year.

Masonry production landscapes are situated along the upper Han River. The trails in the vicinity were damaged by the earthquake and hence were dangerous for community members to use, causing particular inconvenience to elderly people. GIDA actively sought government funds to restore the masonry walls at the headspring of the Han River. The locals reached an agreement to use a participatory design method to renovate embankments on both sides and to restore the unique masonry SEPL. To date, the community has repaired about 3100 m of the destroyed masonry walls, and restored four ha of production areas, with an annual production

Fig. 7.5 Dry masonry construction workshops (Source: NCHU and GIDA)

capacity of 120 tons of persimmon. Gonglaoping has also held several dry masonry construction workshops and invited masonry craftsmen to teach the local youth (Fig. 7.5). The participating community members stated that the theoretical concepts of masonry methods are not difficult to learn; however, considerable effort and skill is required for implementation. Through sharing the experiences and key skills of master craftsmen, the community not only promoted a useful construction method, but also ensured the transmission of knowledge on traditional masonry methods. Doing so ensures that the Gonglaoping community implements low-impact practices for slope stability, achieves resource recycling, maintains environmental capacity and passes on traditional heritage to the next generation.

7.2.3 Enhancement of Biodiversity Conservation

From 1970 to 2000, large amounts of chemical fertilisers were applied to increase yields in agricultural production, bringing cumulative damage to the environment and associated ecosystems. Pesticides and fertilisers were washed into the river and irrigation systems through rain, seriously threatening fish and other aquatic organisms. Excessive use of hazardous chemicals contaminated soil and groundwater, bringing health risks to local farmers and consumers.

Once the community members started to realise the harmful impacts of chemicals on health and the environment, they decided to seek professional assistance. A series of training courses on orchard management was held by experts from COA, including topics such as eco-farming practices, rational application of fertilisers and safety standards for pesticide application (Fig. 7.6). Community residents also worked together with various academia, government agencies, and NGOs to educate locals regarding the long-term impacts of chemical fertilisers and pesticides on soil productivity, farmland ecosystems, harvest quality and human health. Moreover, local residents call for meetings if any related issue needs to be resolved or any assistance is required (Fig. 7.6). This communication mechanism can be seen as the prototype

Fig. 7.6 Training in eco-friendly farming practices (Source: GIDA)

of the future Communications, Education, and Public Awareness (CEPA) strategy that was to be implemented.

The Gonglaoping community began its rural regeneration plan in 2012, funded and supervised by the SWCB. Under the project's operating framework, the community has held nine "Training of Trainers" (TOT) workshops, training locals on organic fertilisers and pesticides, as well as their development and use. Lecturing content included the applications of Trichoderma, a type of biological fertilisation, to treat plant disease and a biocontrol practice for lychee stink bugs involving introduction of *Anastatus japonicus* (wasps). Currently, eco-friendly farming (e.g. persimmon, citrus fruits, and longan) has been expanded to 15 ha, which is about 12.5% of the total production area.

The community set up a water quality monitoring system in 2017 to prevent pollution from agricultural runoff and started making long-term records of observations, particularly examining the streamflow and water quality of the upper, middle and lower streams. Indicators include conductivity, pH value, dissolved oxygen (DO), nitrate concentration, total phosphorus and biochemical oxygen demand (BOD). Various fish and amphibians have been spotted after the monitoring system was put in place, such as the Taiwan striped barb *(Acrossocheilus paradoxus)*, Formosan stripe dace *(Candidia barbata)*, goby *(Rhinogobius rubromaculatus)*, carp *(Carassius auratus)*, temple tree frog *(Kurixalus idiootocus)*, Brauer's tree frog *(Polypedates brauerii)*, Latouchte's frog *(Hylarana latouchii)*, Gunther's frog *(Hylarana guentheri)*, ornate narrow-mouthed toad *(Microhyla fissipes)*, Fujian large-headed frog *(Limnonectes fujianensis)*, and rice field frog *(Fejervarya kawamurai)*.

Starting from the 1960s, locals began to realise the importance of water and soil conservation, and considered stopping the cultivation of crops in the upstream catchment area of the Han River. Afterward, the settlement gradually moved downstream for agricultural activities. Compared to the orchards situated on the terraces, lands in the upstream area required much more effort to grow and manage crops. Besides, farmers were getting older and their suitable farming areas were slowly shrinking. Abandoned lands increased due to fewer disturbances from human activities and hence have turned into secondary forest by natural forest succession. This process created wildlife habitats and made this region abundant with biodiversity. After a 50-year natural succession, 109 ha of natural forest has been established. Field surveys have documented 54 species of plants and 30 avian species.

Additionally, stone masonry embankments on the hillside are constructed by a triangular stacking method, called the *traditional herringbone method*. The stone walls with irregular cobbles have cavities of different sizes, providing spaces and refuge for small animals to hide and plants to grow. The common species include the Taiwan Maesa (*Maesa formosana Mez.*), Taiwan scouring rush (*Equisetum ramosissimum Desf.*), skunk-vine (*Paederia cavalerieri auct.* non H. Lev.), sword brake (*Pteris ensiformis* Burmann), tuberous sword fern (*Nephrolepis brownii (Desv.)* Hovenkamp & Miyam), acuminate leaf morning glory (*Ipomoea indica* (Burm. f.) *Merr.*), five-striped blue-tailed skink *(Plestiodon elegans)*, Swinhoe's japalure (*Diploderma swinhonis*), gossamer-winged butterflies and ants, among others.

7.2.4   Development of a Self-sustaining Economy

Since its establishment, GIDA has actively applied for funding to hold industrial revitalisation activities to attract investors, buyers and tourists. For example, the Citrus Industry and Culture Festival in 2016 had at least 3000 visitors. It increased local incomes by 30% through the sale of agricultural products with revenue of about 300,000 NTD (roughly 10,000 USD). The hotel accommodation rate also went up by 20%. The increased visits and recognition of tourists has enhanced the local youths' confidence in business opportunities and career possibilities in their hometown.

Since 2013, when the first youth returned home, the total number has continued to rise, reaching ten. The young generation of farmers who have moved back to their hometown have insisted on practicing organic and eco-friendly farming, growing citrus fruits and persimmons with the aim of protecting natural resources and the environment. Although mature fruits (e.g. citrus) were not attractive looking and could not sell at a good price during the transition period, local famers unleashed their creativity to make a wise turn. Instead of making money by selling traditional table fruit, they developed two post-harvest processed products, organic persimmon leaf tea and organic citrus gummy candies (Fig. 7.7), that could earn more profit. Both products are made from local persimmons and citrus fruits that are cultivated using organic farming methods. The tender persimmon leaves that grow in spring are carefully selected for processing. Persimmon leaves are highly nutritious and rich in protein, amino acids and multi-vitamins. The tea is also acknowledged as one of Rural Goodies" in Taiwan Island (i.e. good products from rural communities) by the SWCB, generating a revenue of up to 370,000 NTD (12,300 USD) annually. In addition, the organic citrus gummy is a soft candy made from local tangerine oranges. It is rated as one of the ten best souvenirs in Taichung by the media and

Fig. 7.7 Organic agricultural products of the Gonglaoping community (Source: GIDA)

can now be bought on the Taichung city government's e-commerce platform, generating annual revenue of 660,000 NTD (21,900 USD).

Under the leadership of the youthful chairman of GIDA, the Gonglaoping community has begun to offer a variety of lively activities. The community operates four in-depth agricultural tours based on the four seasons (i.e. spring, summer, fall, and winter) so that people from all over the world can better understand the local characteristics of Gonglaoping. Local industrial development and promotional events include "Spring Tour in Gonglaoping - Orange Blossom Viewing and Citrus Essential Oil DIY", "Lovely Lychee in Gonglaoping - Fengyuan Lychee Festival and Outing", "Autumn Praise: Gonglaoping Persimmon Banquet and Rural Chinese Orchestra Concert", and "Winter-When-Tangerines-Turn Red--Gonglaoping Picnic Party."

7.3   Results

By engaging in cross-sector collaboration with multiple stakeholders, including SWCB, the National Chung Hsing University (NCHU) and other local groups, the following achievements have been made:

(1) **Resolving conflicts surrounding water use by increasing water storage capacity and reducing groundwater consumption during the dry season**
    Rainfall patterns in Gonglaoping are uneven by seasons. The rainy season is in summer while the dry season is in winter. Both the persimmon growing period (summer) and the dry season (winter) require a lot of water for irrigation. However, the Han River cannot provide an adequate amount of water; thus, farmers often solved the problem by pumping groundwater, which affects the water and groundwater usage of residents living midstream and downstream. The community carried out an expansion of the original water retention capacity and also built terraced-type, ecological engineered reservoirs by consulting experts in SWCB and NCHU. The total volume of reservoirs was expanded from the original 56 tons to 416 tons. Based on an estimation, if the reservoirs are filled to capacity during the rainy season, they would able to match water

Fig. 7.8 Community development meetings have been held every Tuesday night for 15 consecutive years (Source: GIDA)

demand during the dry months (February and March), reducing the groundwater used to an amount equivalent to that used for 5 weeks. Accordingly, the groundwater extraction volume was reduced from the original 510 to 93 tons. On the other hand, irrigated areas that are supplied by the Han River increased from 10 ha to 14 ha (17.5% of the catchment area) and persimmon production also increased by about 120 tons. This construction was a win-win strategy to alleviate the problem of insufficient irrigation and to conserve wildlife habitats.

(2) **Restoring traditional masonry in the SEPL**

Hillside orchards represent a unique SEPL with multi-layer stone masonry embankments. The embankments conserve water, stabilise slopes and provide refuge for various organisms. The holes and cavities between the embankments are home to a variety of small creatures. Vegetated buffer strip cultivation also benefits the environment in many ways, including reducing herbicide use, alleviating soil erosion, balancing soil moisture, increasing soil nutrients and maintaining healthy soil ecosystems.

(3) **Empowering, enhancing and sustaining local operations through the leadership of a community-led NGO, the Gonglaoping Industrial Development Association**

The Gonglaoping Industry Development Association, which was established in 2004, plays a vital role in sustainable water resources management. The association has experienced seven directors. All previous directors continue to participate in community industrial development. The association meets regularly every Tuesday night to discuss development affairs and to bring all locals together to promote unity in industrial revitalisation (Fig. 7.8). For the protection and rational use of the Han River's water resources, the association plays a role in the coordination of the sustainable management and utilisation of water resources. To mitigate conflicts in the processes of irrigation, community members reached an agreement to set up a neighbourhood watch to patrol the Han

River and drafted a convention to conserve the water resources of the Han River and manage the community.

(4) **Enhancing benefits of green industries: the new economic and ecological values of organic industries**

The youth have developed organic products such as persimmon leaf tea and citrus gummy candies to add new economic value to local industries. In particular, they have adapted innovative post-harvest processing technologies to make new healthy foods. The persimmon leaf tea can earn twice the profits of sale of the traditional agricultural product, while the revenue from citrus gummy candies is even better with a boost of ten times the profit. Because both products emphasise organic concepts and zero fertiliser residue, the success of these products provides solid evidence on the benefits of organic farming practices, brings new insights and motivates transformative change in climate smart agricultural practices. The eco-friendly farming practices adopted by the community represent a true win-win strategy to maintain water quality, conserve biodiversity and sustain the income of local people.

The cumulative number of visitors in 2017 was about 1,450, with an estimated 25% return rate and the net economic benefit of 1.27 million NTD (421,000 USD). This implies that the innovations in the community activities in Gonglaoping are able to constantly attract customised groups.

7.4 Discussion

The development of foothills often requires slope stabilisation. The dry masonry piled up with *in situ* materials not only solves the problem of stone placement after site preparation, but also has good water permeability and generates habitats for fauna and flora. The masonry SEPL presented has many valuable functions, including soil and water conservation, restoration of the practice of low–impact masonry, maintenance of biodiversity, creation of refuge for organisms, integrated applications of traditional knowledge, and promotion of tourism.

Constrained by micro-climatic conditions, the Gonglaoping community is often short of water for irrigation in dry seasons. Fortunately, most community members are aware of and understand particular issues; they show empathy towards each other. Based upon mutual trust, locals have established a strategic, decision-making routine on a weekly basis. Their scheme divides water allocation by divisions and time slots, so that all orchards can be irrigated to maintain vigorous growth. Also, the water allocation plan is reviewed periodically and the need for revision determined. This helps avoid conflicts caused by competing water sources and mitigates the over-pumping of groundwater in the dry season. Hence, this communication mechanism is an effective strategy that provides incentives to reach consensus.

The Gonglaoping community is situated on the upper Han River. Any over-use of fertilisers and pesticides could impact negatively on water quality and risk community health. Moreover, the irrigation relies on river water supply, leading to a vicious cycle. The farmers understand that high quality fruits can only be produced in healthy ecosystems with good water quality. Thus, in rejecting the use of herbicides and adopting eco-friendly farming practices, the community has been able to maintain the ecological environment and water quality near the Han River. Several local farmers have cooperated with government agencies to apply Trichoderma instead of pesticides and chemical fertilisers, and also perform lychee stink bug biocontrol by introducing *Anastatus japonicus* (wasps).

In the process of post-disaster recovery and restoration, the community created numerous job opportunities in the field of leisure agriculture and has hosted a variety of activities and events to lure visitors from urban areas. These opportunities have been able to attract young people to move back to the rural areas. Besides, the ideas proposed by these returning youths have been accepted by the community, and they participate in local affairs, hence becoming more deeply rooted in the community affairs. In 2016, the chairman of GIDA was elected. The community believes that passing down authority to the younger generation opens up a new era and broadens the path to future prosperity. Thus, they voted for a youth to serve as the new chairman. The new chairman and other returning youths have been led by the former president, which has helped them to quickly become familiar with community affairs. Also, they have altered the traditional way of organising events by integrating online marketing, social media and other trendy concepts. This makes the activities more dynamic and shines a light on the Gonglaoping community's features.

GIDA has shown a strong connection with locals, and actively proposes improvement plans while acting as a "glue" agent and cooperating closely with various governmental agencies, the private sector, and other NGOs. This multi-stakeholder platform enables transformative change and enhances related movements toward sustainability in SEPLs. Current associated stakeholders include the Taichung Branch of SWCB, Agricultural Bureau of Taichung City Government, NCHU, Fengyuan District Farmers' Association, Agricultural Research and Extension Station, and the Taiwan Agricultural Chemicals and Toxic Substances Research Institute. Although the Gonglaoping community still needs external funding (e.g. governmental incentives or competitive grants), its common goal is to sustain economic revenue and ultimately reach a self-sustaining status in the future.

7.5   Conclusions: Key Messages

Achieving the sustainability of SEPLs not only requires the application of a holistic approach, but also cross-sector collaboration (community-government-university). The key messages of this case are:

(1) The masonry SEPL highlights many valuable functions, including soil and water conservation, a low-impact practice for slope stabilisation, maintenance of biodiversity, preservation of traditional wisdom and tourism.
(2) Constrained by micro-climatic conditions, the Gonglaoping community is often short of water supply for irrigation in the dry seasons. Locals have learned to facilitate a strategic decision-making routine, with the water allocation plan reviewed and revised weekly so that all orchards can be irrigated for growth. In addition, this mechanism helps avoid conflicts of water use and halts overuse of groundwater in dry seasons. The key element of such effective operation is mutual trust among community residents.
(3) Because of the success of organic products, locals are more willing to practice eco-friendly farming, which is a true win-win strategy to sustain water quality, biodiversity and the economy.

Hence, we are confident that the Gonglaoping community could serve as an example for practical implementation of the Satoyama Initiative on hillside, agricultural landscapes in East Asia.

Acknowledgements Sincere gratitude and thanks are expressed to local residents, farmers, and different groups of the Gonglaoping community for providing valuable assistance for the study. We have prepared associated contents as an IPSI case study, which we are planning to submit to the IPSI Secretariat. The work was supported by Rural Regeneration fund of the Soil and Water Conservation Bureau, Council of Agriculture, Executive Yuan, R.O.C. (Chinese Taipei).

References

Bélair, C., Ichikawa, K., Wong, B. Y. L., & Mulongoy, K. J. (Eds.). (2010). *CBD Technical Series: Sustainable use of biological diversity in socio-ecological production landscapes: Background to the 'Satoyama Initiative for the benefit of biodiversity and human well-being.* Montreal: Secretariat of the Convention on Biological Diversity, Technical Series no. 52184.

IPBES. (2019). Global assessment report on biodiversity and ecosystem services of the Intergovernmental Science-Policy Platform on Biodiversity and Ecosystem Services. In S. Díaz, J. Settele, E. S. Brondízio, H. T. Ngo, M. Guèze, J. Agard, A. Arneth, P. Balvanera, K. A. Brauman, S. H. M. Butchart, K. M. A. Chan, L. A. Garibaldi, K. Ichii, J. Liu, S. M. Subramanian, G. F. Midgley, P. Miloslavich, Z. Molnár, D. Obura, A. Pfaff, S. Polasky, A. Purvis, J. Razzaque, B. Reyers, R. R. Chowdhury, Y. J. Shin, I. J. Visseren-Hamakers, K. J. Willis, & C. N. Zayas (Eds.), *IPBES Secretariat.* Bonn: IPBES.

Natori, Y., Dublin, D., Takahashi, Y., & Lopez-Casero, F. (2018). *SEPLS: Socio-ecological production landscapes and seascapes—experiences overcoming barriers from around the world.* Tokyo: Conservation International Japan.

SWCB Taichung Branch, COA., R.O.C. (Chinese Taipei). (2017). *2017 rural community intelligent disaster prevention and Satoyama initiative practice in hillside area.* Taichung City: SWCB Taichung Branch, COA.

SWCB Taichung Branch, COA., R.O.C. (Chinese Taipei). (2018). *2018 strategy development and implementation of the Satoyama initiative in Chung-Miao rural communities.* Taichung City: SWCB Taichung Branch, COA.

Chapter 8
Transformative Change in Peri-Urban SEPLS and Green Infrastructure Strategies: An Analysis from the Local to the Regional Scales in Galicia (NW Spain)

Emilio Díaz-Varela, Guillermina Fernández-Villar, and Alvaro Diego-Fuentes

Abstract Transformative change involves the integration of different social dimensions and the involvement of a multiplicity of actors resulting in high levels of complexity. Considering all this, our work addresses the development of green infrastructure (GI) to improve the conservation of biodiversity and the provision of ecosystem services from two different approaches and scales: regional and local.

From the regional level, a GI strategy was promoted by the regional government of Galicia (NW Spain) through institutional efforts following a multidisciplinary approach including public participation processes. On the other hand, a local, participative perspective is exemplified in the Neighbourhood Association of the Parish of Chapela (Redondela, Galicia), a peri-urban, coastal area where intensive forestry and urban expansion threatens the availability of accessible multifunctional ecosystems for the local communities.

Both approaches are indicative of seeds for a transformative change yet to happen. Nevertheless, they differ in their visions, values and goals: the regional level is statutory-oriented and focused on the accomplishment of administrative objectives; the local level is based on the communities' wellbeing aims and calls-for-action. Differences are also detected in the risks and barriers to transformative processes, from the inertia of administrative procedures to the limitations of local action to face environmental and developmental problems. Exploration of these contrasting perspectives leads to the identification of needs for institutional change, the emergence of new governance systems, and the development of new perspectives for strategic planning and management.

E. Díaz-Varela (✉) · G. Fernández-Villar
Higher Polytechnic School of Engineering, University of Santiago de Compostela, Lugo, Spain
e-mail: Emilio.diaz@usc.es

A. Diego-Fuentes
Community Facilitation & Environmental Education - Neighbourhood Association of Chapela, Pontevedra, Spain

© The Author(s) 2021
M. Nishi et al. (eds.), *Fostering Transformative Change for Sustainability in the Context of Socio-Ecological Production Landscapes and Seascapes (SEPLS)*,
https://doi.org/10.1007/978-981-33-6761-6_8

133

Keywords Green infrastructure · Strategic planning · Ecosystem services · Common land · Ecosystem management · Transformative change

8.1 Introduction

8.1.1 The Different Levels of Transformative Change

The increasing impact of human activities on ecosystems and the general decline of life on Earth, well documented in the recent Intergovernmental Science-Policy Platform on Biodiversity and Ecosystem Services (IPBES) Global Assessment (IPBES 2019a, b), call for transformative change to halt the future trajectories of the continuous decline of living systems and their related contributions to people (McAlpine et al. 2015; Díaz et al. 2019). Such transformative change depends on the application of governance mechanisms ('levers') on the priority intervention 'leverage' points described in Chap. 1 of this SITR volume. Specifically, in the *nexus approach* to achieve the SDGs described in the aforementioned IPBES report, the aim of "sustaining cities while maintaining the underpinning ecosystems (both local and regional) and their biodiversity" is included (Chan et al. 2019, p. 8), addressing specifically goals 11 (Sustainable cities and communities) and 15 (Life on land) of the SDGs. City-specific targets, including retention of species and ecosystems and limits on urban transformation, are to be achieved by strengthening local- and landscape-level governance and enabling transdisciplinary planning to bridge sectors and departments, and to engage businesses and other organisations in protecting public goods. This focus integrates the achievement of biodiversity conservation objectives with those geared to improvement of local quality of life. In this sense, the integration of ecological ('green') and built ('grey') infrastructure are considered as increasingly important (Chan et al. 2019, p. 8), being the design and maintenance of ecological connectivity across the territory, and specifically in the interfaces between urban and rural areas, critical for both people and nature.

In alignment with these objectives, the European Union in its Biodiversity Strategy to 2020 states its Target 2 as follows: "By 2020, ecosystems and their services are maintained and enhanced by establishing green infrastructure and restoring at least 15% of degraded ecosystems" (European Commission 2011, p. 12). To this end, a strategy for green infrastructure was developed (European Commission 2013a). In it, green infrastructure (GI) is defined as

> ...a strategically planned network of natural and semi-natural areas with other environmental features designed and managed to deliver a wide range of ecosystem services. It incorporates green spaces (or blue if aquatic ecosystems are concerned) and other physical features in terrestrial (including coastal) and marine areas. On land, GI is present in rural and urban settings (European Commission 2013a, p. 3).

The vision is essentially multifunctional (DG Environment 2012), and is considered as a successfully tested tool for providing ecological, economic and social benefits through natural solutions, contributing to comprehension of the value of nature benefits to society and to mobilising investments to sustain and reinforce them (European Commission 2013a). The strategy should be implemented at the national level by the EU state members, which in turn would coordinate the implementation at the regional level. Thus, in Spain, the development of a "State-level Strategy for Green Infrastructure and Ecological Connectivity and Restoration" was approved (eds. Valladares et al. 2017), being the framework to carry out regional-level strategies by the different autonomous communities. One of the autonomous communities pioneering the development of a GI strategy was Galicia.

In Galicia, despite deep structural and functional changes suffered by rural areas, cultural landscapes still persist as socio-ecological production landscapes and seascapes (SEPLS) in many rural areas (Calvo-Iglesias et al. 2006, 2009; Morán-Ordóñez et al. 2011). Nevertheless, there are still examples of SEPLS in peri-urban areas continuously evolving due to the influence of diverse and complex driving forces and interrelationships: urban development, urban-rural interaction, peri-urban communal land dynamics, etc. (Souto-González 1993; Swagemakers and Dominguez-García 2015; Dominguez-Garcia et al. 2015). The local communities living in these geographical areas participate in their management espousing different visions, values and perspectives, from productive activities to environmental conservation. How may these activities lead to transformative change at the local level? And how does this relate to the top-down implementation of green infrastructure strategies at the regional level?

8.1.2 Objectives

We aim to compare two different approaches at the regional and local level under the same research question: How is the development of green infrastructure elements achieved in order to improve the conservation of biodiversity and the provision of ecosystem services? The first, developed by the regional administration, provides a statutory framework for action. The second, developed by a local "Neighbourhood Association" in Chapela (Redondela, Spain), exemplifies the emergence of new values in local communities and institutions leading to the demand for a better environment. The analysis of both perspectives allows for a systemic approach to the governance system, its elements and interactions, leading to the acknowledgement of not only how transformative processes may emerge, but also of the limitations, risks and barriers involved in changes yet to happen in peri-urban SEPLS.

8.2 Material and Methods

8.2.1 Study Area

The approach taken in this work addresses two geographical levels: regional and local (Figs. 8.1 and 8.2 and Table 8.1). The regional level (Fig. 8.1) is represented by the Autonomous Community of Galicia, NW Spain (Framed between latitudes 43° 48′ N and 41° 49′ N and longitudes 6° 44′ W and 9° 18′ W). It spans 29,577 km² and has 2,698,764 inhabitants (INE 2019). The region is located in the Atlantic biogeographical region and is characterised by an oceanic temperate climate, with zones of Mediterranean climate influences. Land uses are occupied by artificial surfaces (6.52%), surface waters (0.79%), crops (24.56%), forestry areas (17.12%), native forests (14.98%), shrubland and pastures (31.21%) and other ecosystems (4.81%) (SIOSE 2011). The region is to a large extent rural, with some concentration of industrial activity in the metropolitan areas of A Coruña, Ferrol and Vigo.

The local level (Fig. 8.2) is represented by a peri-urban coastal area near the aforementioned city of Vigo. The area was defined by the basins of four small streams (Cabras, Fondón, Maceiras and Pugariño) flowing E-W through the rural-urban gradient. The main settlements are Cidadelle, Angorén, Laredo, A Igrexa and Parada, in the parish of St. Fausto of Chapela; Trasmañó, Igrexa and Cabanas in

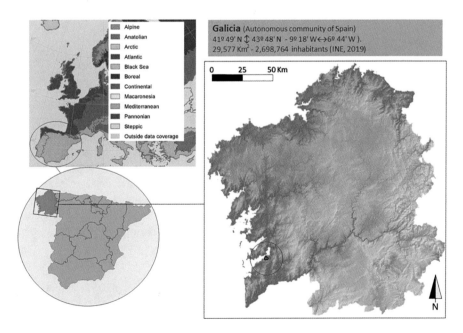

Fig. 8.1 Study area: regional level (Source: prepared by authors; digital terrain model and other GIS data from Centro Nacional de Información Geográfica (2019); biogeographical map from European Environment Agency (2017))

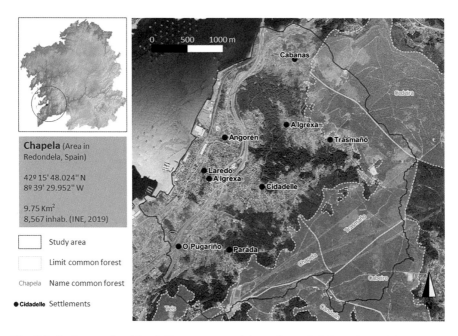

Fig. 8.2 Study area: local level (Source: prepared by authors; digital terrain model and aerial imagery from Centro Nacional de Información Geográfica (2019))

Table 8.1 Basic information of the study area

Country	Spain
Province	Galicia
District	Galicia
Municipality	n.a.
Size of geographical area (hectare)	2,957,700
Number of indirect beneficiaries	2,698,764
Dominant ethnicity(ies), if appropriate	Caucasian
Size of case study/project area (hectare)	975
Number of direct beneficiaries	8,567
Dominant ethnicity in the project area	Caucasians
Geographic coordinates (latitude, longitude)	42° 15' 47.88" N; 8° 39' 29.88" W

St. Vicente of Trasmañó; and O Pugariño, in St. Salvador of Teis. Parishes are historical, non-statutory territorial units still in use for local organisational purposes. The former two are in the municipality of Redondela, and the latter, in the municipality of Vigo. The total population is 8,567 inhabitants (INE 2019). The area shows a close interface between urban (residential and industrial) uses and forestry areas (mainly comprised of *Eucalyptus globulus* plantations and shrubland). Both intensive land uses put pressure on riparian and forest ecosystems (which include the *Prunus lusitanica* species of tree, classified by IUCN as "vulnerable") that local communities wish to preserve. Forest areas still involve ownership regimes known as "Communal Forest Land" ("*Montes Veciñais en Man Común*" (MVMC) in Galician language). The MVMC are community managed, and all decisions are made democratically through the assembly of the Community of MVMC (CMVMC), with a board acting as representative body. Generally, each MVMC corresponds to a parish or settlement, and the main ones in the area are those of Chapela, Teis, Candeán, Cabeiro, Trasmañó and Cedeira (Fig. 8.2).

8.2.2   Methodological Approach

8.2.2.1   Document Analysis

Documents relevant to the study case were reviewed in order to identify keywords and specific contents related to the views, values and perspectives needed for transformative change. Such documents included the current regulations at the European level regarding biodiversity conservation and green infrastructure (European Commission 2011, 2013a). Also, at the national level, the documents regarding the next Green Infrastructure Strategy (eds. Valladares et al. 2017) were reviewed. At the regional level, the Galician GI Strategy website (Infraestructura Verde de Galicia 2019), as well as the unpublished documents from the team commissioned for the development of the Strategy (see next section), were included. Finally, some of the documents prepared by the Neighbor Association of Chapela were reviewed, including their transversal programme for environmental education and heritage (Diego Fuentes 2019), and their proposal for the improvement of the municipal Sustainable Urban Mobility Plan (PMUS) (Neighbourhood Association of Chapela 2019).

8.2.2.2   Spatial Analysis

Spatial data, including vectorial and raster GIS and remote sensing data from the National Geographic Institute of Spain (Centro Nacional de Información Geográfica 2019), as well as the Spatial Data Infrastructure from Galicia (Xunta de Galicia 2019) was used to analyse the temporal evolution of the SEPLS, alongside the proposals made by the Neighbourhood Association of Chapela.

8.2.2.3   Direct Observation

The direct participation of all the authors in the different levels addressed in this work allowed for direct observation of developments and outcomes. One of the co-authors, Alvaro Diego Fuentes, is the environment delegate of the Neighbourhood Association of Chapela. The lead-author, Emilio Díaz-Varela, participated in the development of the Galician GI Strategy as one of the interdisciplinary team members. Finally, the other co-author, Guillermina Fernandez-Villar, developed an academic work (Fernandez-Villar 2019), analysing the proposals of the Neighbourhood Association of Chapela from the point of view of green infrastructure implementation. The information retrieved as a result of direct observation was the source used to develop the indicators (see Sect. 8.2.2.4).

8.2.2.4 Indicators

In order to characterise the differences between the two analysed approaches, we developed a set of indicators specific for this work. These indicators allow for summarising and conclusions on both perspectives. Due to the character of the information to be retrieved and to the methodological approach, all the indicators used are of qualitative and descriptive character. They are described as follows:

- *Vision*: The general purpose envisaged for the GI
- *Triggers*: The enabling factors for GI development
- *Approach*: The general view for implementation of the GI
- *Coordination*: How the institutional and other actors involved or interested in the GI would be coordinated for implementation
- *Aims*: Specific objectives and implementation levels of the GI
- *Means*: How the aims would be put into practice
- *Identified barriers*: How the current configuration of the governance system could hinder implementation of GI supportive of transformative change
- *Identified needs*: Elements to overcome the *identified barriers*

8.3 Results: Two-Level Approach Towards Green Infrastructure

8.3.1 The Regional Level: Statutory Approach

The Galician GI Strategy responds to the application of European-level directives related to biodiversity and green infrastructure (European Commission 2011, 2013a), and is coordinated with the state-level GI Strategy (See Fig. 8.3b). In this sense, it responds to the global demands of society for environmental conservation (See Fig. 8.3a), being coherent with the city-specific targets described in the IPBES Global Report for transformative change (Chan et al. 2019, p. 8). For the development of the Galician GI Strategy, the (then) Regional Ministry for Environment and Land Use Planning (CMAOT) commissioned a multidisciplinary team coordinated by the Institute of Territory Studies (IET) and composed of three research groups of the University of Santiago de Compostela (USC) and two from the University of A Coruña (UDC). Later, a private company specialising in sociological studies

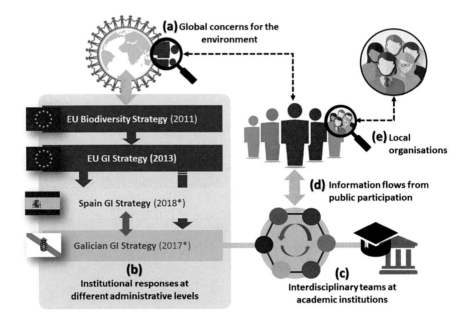

Fig. 8.3 Governance chart for the development of Green Infrastructure at different scales. *Note*: (**a**) Global concern in societies and institutions demands action for conservation of biodiversity and ecosystem services. (**b**) Organisations at different levels respond with regulations focused in GI Strategies (among others). (**c**) At the regional level, interdisciplinary groups assume the collaboration between regional government and academia to develop GI strategies. (**d**) Public participation processes are developed to integrate the needs and visions of the population. (**e**) Local organisations at local levels contribute their visions and needs

(DELOGA) joined the team for the development and organisation of the public participation (sub-) strategy. The approach was then statutory, combining a top-down design with the integration of bottom-up suggestions and information coming from the public participation sub-strategy. The agreement for the multidisciplinary team was signed in August 2017, and the project was planned to be developed in 16 months.

The methodological approach was structured in seven working packages (WPs): WP0, to be developed by IET, was project management and coordination. WP1 (Biodiversity Mapping), WP2 (Ecosystem Service Mapping), WP3 (Delineation of Green Infrastructure Elements), WP4 (Socio-economic Implementation), and WP5 (Formalisation of Directives and Specific Strategies) were developed in a coordinated effort by the USC and UDC teams, with the support of IET. Finally, WP6 (Public Participation) was developed by DELOGA, but USC, UDC and IET collaborated on the participative events.

Figure 8.4 summarises the Galician GI Strategy in two graphic examples. The first (Fig. 8.4a) shows the network of green infrastructure elements, including core and buffer areas, landscape and structural corridors, and multifunctional zones. The second (Fig. 8.4b) classifies Galicia in sub-regional areas, where three different

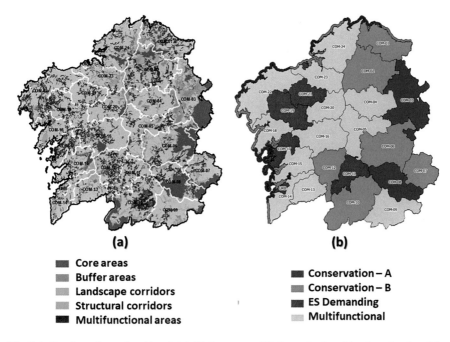

Core areas
Buffer areas
Landscape corridors
Structural corridors
Multifunctional areas

Conservation – A
Conservation – B
ES Demanding
Multifunctional

Fig. 8.4 Results at the regional level: (**a**) GI elements and their typologies; (**b**) sub-regional social-ecological units. See text for details. (Source: Modified from Díaz-Varela et al. 2018)

sub-strategies for socioeconomic improvement related to GI and ecosystem services were developed. These were adapted to the social-ecological demands of each area, and integrated the results of the public participation: conservation (addressed to sub-regional areas with more (A) or less (B) core areas), multifunctional (sub-regional areas with a combination of different GI elements), and ecosystem service demanding (sub-regional areas with scarce GI elements).

8.3.2 The Local Level: Community Approach

The evolution of the SEPLS in the area of Chapela (see Sect. 8.2.1) is important for the comprehension of current dynamics. The analysis of aerial photographs from 1946 to 2017 (Fig. 8.5a) shows a rapid decline in the social-ecological system due to the expansion of urban fabric and linear infrastructure. Such changes degraded the system and the strong relationships between settlements, 'infield' terraced structures, and 'outfield' productive shrubland (Bouhier 2001; see Fig. 8.5b, 1-2-3). Progressive non-planned, spontaneous urbanisation used former structures to build houses and roads (Fig. 8.5c, 2-3), and the breakage of the relationship between 'infield' and 'outfield' induced afforestation in the former (Fig. 8.5c, 1). Newcomers with little relationship with the former social-ecological system also altered the composition

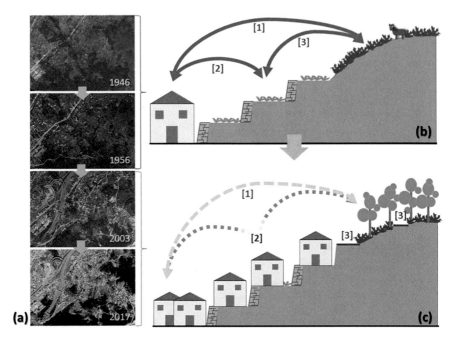

Fig. 8.5 Evolution of the social-ecological system: (**a**) Aerial photographs of part of the study area taken in 1946, 1956, 2003 and 2017 show the changes that have taken place in the system. (**b**) Schematic interpretation of the social-ecological system towards the 1950s; and (**c**) towards the 2020s. See text for details

and function of the social fabric: a new SEPLS was formed, with novel needs and expectations regarding green infrastructure, derived from specific perspectives and values. Thus, instrumental values centred in the productivity of the territory have lost relevance (even while basic forestry activities remain), while new relational values have been revealed through new needs for the community's environmental quality, which in turn underpin management perspectives and sustainability visions for the SEPLS.

It is in this context that the Neighbourhood Association of Chapela was developed. This is a non-governmental, non-profit association founded to defend the interests and needs related to the wellbeing of the community, including the improvement of its environment. To this end, the association makes requests to local and regional administrations for interventions in the environmental improvement of the neighbourhood. These are normally accompanied by well-developed reports and recommendations of a strategic character. Examples range from contributions to the Sustainable Urban Mobility Plan (PMUS) and the General Municipality Urban Plan (PXOM) at the local level (Fig. 8.6), to the Galician Strategy of Circular Economy, and also the Galician Green Infrastructure Strategy discussed herein (Fig. 8.4). The proposals are also oriented to generation of environment-related employment, a social component that may be crucial for attitude changes.

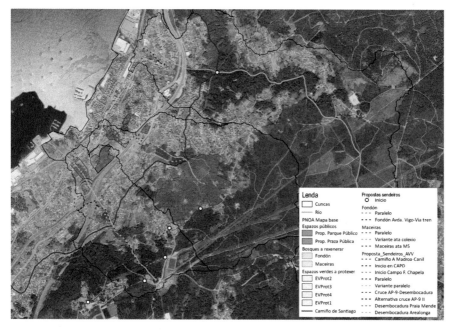

Fig. 8.6 Example of the spatial planning proposals made by the Neighbourhood Association of Chapela. The map reflects the community's vision to overcome the environmental restrictions of the area through development of site-level greenways, corridors and open spaces, establishing the desired green infrastructure (Source: Neighbourhood Association of Chapela 2019; Fernandez-Villar 2019)

On the other hand, the association also is highly involved in educational projects, oriented toward transformation and the encouragement of participatory citizenship. These include talks and workshops in schools to disseminate information on waste management, programmed obsolescence and the encouragement of circular economy with a holistic approach. Students are taught that the right to a good environment and to sustainable development are a part of human rights. Also, participatory activities, such as excursions to clean up waste in local riverbeds, are conceived as educational events where students are stimulated to reflect on the consequences of social and environmental activities in production and consumption processes (Fig. 8.7). The design of this educational approach is strongly based on social psychology and social change theories, and aims to be of a transformative character.

Fig. 8.7 Participatory excursion to clean up waste in riverbeds and associated heritage elements. The activities are designed to create an affective link with the environment

8.4 Implications for Transformative Change: Visions, Values. . . and Barriers

8.4.1 The Regional Level

The initial agreement for the development of the Galician GI Strategy (Xunta de Galicia 2017, p. 5) is oriented towards an institutional instrument integrative for land planning, green infrastructure development, and functional connections between landscapes and territories. These aims show an alignment with the visions behind EU and state directives, specifically with the integrative orientation of the EU GI Strategy, which aims to make a significant contribution to regional development, climate change, disaster risk management, agriculture/forestry and the environment (European Commission 2013a).

The vision is, thus, a "smart solution for today's needs" (European Commission 2013b) integrating conservation of natural capital and biodiversity, enhancement of ecosystem services provision, and economic development, trying to respond to the concerns in societies and institutions regarding such, as well as the basic components of wellbeing (see Fig. 8.3a, b). Nevertheless, a strategic spatial planning component to the GI was included at the regional level. While this is a desirable aspect in the implementation of green infrastructures (Lennon and Scott 2014; Grădinaru and Hersperger 2019), current approaches to such integration are considered insufficient (Ronchi et al. 2020), and the need for new integrative, adaptive and participative approaches is indicated, implying complete institutional change at multiple levels of governance, including administrative bodies, competence of practitioners and capacity building for public participation (Botequilha-Leitão and Díaz-Varela 2020). For instance, in Galicia, the spatial planning system presents coordination problems between different levels and sectoral approaches (Lois-González and Aldrey-Vázquez 2010; Tubío-Sánchez and Crecente-Maseda 2016). Also, institutional barriers and short-term vision have affected the regional GI Strategy. Namely, changes in governance (Decree 42/2019) during the final stages of its development modified the institutional staff in charge of the strategy, resulting in changes in priorities and visions, and a delay in the stages crucial for implementation of the strategy. Consequently, transformative change must involve not only planning and institutional systems, but also innovative governance approaches that need to be integrative, inclusive, informed and adaptive (Díaz et al. 2019). These approaches should catalyse change through multilevel/multiscale learning, connecting through the diverse boundaries of knowledge, values, levels and organisations (Granberg et al. 2019).

8.4.2 The Local Level

At the local level, visions are largely developed based on immediate perception of the environment by local inhabitants. In fact, one important trigger in Chapela was the restricted mobility sensed by the community: non-planned urban structures constrained pedestrian movement, as current streets and roads were built by paving pre-existent pathways more suitable for vehicular traffic, neglecting walkways. Paradoxically, this progressively hindered both access to public transportation and pedestrian mobility. The development of a complex network of highways and railways in the last decades (Fig. 8.5a) aggravated the problem. In a search for solutions, the Neighbourhood Association of Chapela made proposals for reactivating pedestrian communications throughout the peri-urban fabric, and also access to the common forests up in the hills (Fig. 8.6). Interestingly, as some of the old paths follow streams and brooks, efforts made for their reclamation led to an increased awareness of the associated natural and cultural heritage. The involvement of the association in the clean-up of riverbeds, where waste had accumulated over decades of dumping—another consequence of the lack of urban planning (Fig. 8.7)—contributed to a reappropriation of the environment by the community, including the discovery of traditional watermills in the area as important cultural heritage elements, leading to serious efforts for their recovery. The activities of the association are thus intended to provide an educational experience, designed with a strong basis in social psychology. Specifically, collective events for clean-up of riverbeds are aimed at effecting changes in attitudes and behaviour through participation. The resulting emotional involvement has proved to be in this sense more effective than analytical thinking (Small et al. 2007). The consequent participation involves the valuation and appropriation of the environment by the participants (Cooney 2010), and an increasing understanding of the former social-ecological system. In this sense, these activities promote elements of transformational change

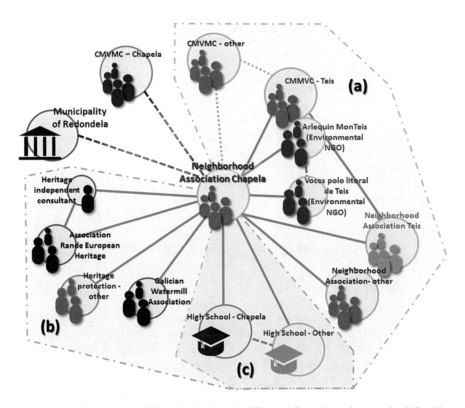

Fig. 8.8 Stakeholder map of Chapela showing the different 'cliques' or sub-networks defined by stronger thematic ties: (**a**) environmental cooperation clique; (**b**) cultural heritage cooperation clique; (**c**) educational clique. Green connections: active; red-green: to be improved; grey discontinuous: to be consolidated. Degree of discontinuity in lines is proportional to the strength of the ties

(McAlpine et al. 2015), involving ethical responsibility in dealing with people and environment, community integration, and reconnecting with and valuing nature.

The strong relationships held by the association with other local organisations (Fig. 8.8) are also an important element for transformative change, and necessary for effective adaptation when transformative change goes beyond administrative and territorial boundaries (Granberg et al. 2019). Thus, environment-related activities are carried out with other neighbour associations and environmental NGOs (Fig. 8.8a), as well as with adjacent communal forests (CMVMCs; see Sect. 8.2.1). Collaborative educational activities are also carried out with the high school of Chapela (and others). Still, the association holds divergent views from those of the CMVMC of Chapela concerning the management of the communal forest. Currently, this CMVMC has delegated management duties to a private company (a possible symptom of institutional apathy), putting focus on the production of fast-growing species for timber. The vision represented by the association, though, is more related to the multifunctional use of the forest and the generation of employment via projects, including edible forests or permaculture. Nevertheless, new negotiations are planned in search of common ground. Finally, the association also interacts with different levels or government, mainly at the local level, but also at the regional level. Nevertheless, some of the association's suggestions for environmental initiatives have come up against barriers. As local administration, the municipality has been reactive to initiatives that are unusual in their strategic, collaborative implementation. Yet, when the association invited municipality representatives to the area to see in situ the current state and motivations for the actions, some changes in the municipality's visions and attitudes took place. The association also took part in public participation sessions for the development of the Galicia GI Strategy, with their initiatives included as examples of potential local-based actions by NGOs.

8.4.3   Local vs. Regional Approaches: Indicators and Lessons Learned

Table 8.2 shows a comparison between the two analysed levels, based on the set of indicators described in Sect. 8.2.2.4. The descriptions associated with each indicator are related to the respective contributions of the actions for implementation of GI at each level to transformational change.

It is possible to identify a convergence in both approaches towards the improvement of environmental conditions in the territory, via the implementation of actions through GI. Nevertheless, clear differences can also be detected, associated with their top-down and bottom-up characteristics. Thus, for instance, the regional level pursues integrative territorial planning triggered by the society's perception of environmental degradation, while at the local level the focus is on the improvement of the neighbouring environment based on local perception of mobility limitations and environmental quality. In a similar way, GI-based solutions are adopted through institutional actions at the regional level, whereas at the local level community initiatives and network coordination are much more relevant. Consequently, the identified needs and barriers are consistent with these differences in the scale-based approaches.

As discussed in Chap. 1 of this volume, enabling transformative change implies implementation of priority interventions, or "levers", targeting key points of intervention, or "leverage points". Effective use of levers and leverage points requires innovative governance approaches and organising the process around nexuses (Díaz et al. 2019). Thus, on the one hand, new models of governance should target the integration of multidirectional information (including iterative learning loops) and connections at different levels. On the other hand, nexuses represent closely interdependent and complementary goals, and reflect interactions between multiple sectors and objectives. In this sense, implementation of GI strategies, a basic element in the nexus approach for "sustaining cities while maintaining the underpinning ecosystems and their biodiversity", becomes a tool for achieving interlinked goals. In this work, the lessons learned for transformative change are related to both

Table 8.2 Detailed comparison between regional and local levels defined for the set of indicators

Indicators	Regional level	Local level
Vision	Integrative territorial planning to create functional connections between environmental planned spaces	Improvement of local environmental quality through multifunctional concepts combining natural and cultural heritage
Triggers	International/global perception of environmental degradation approached regionally through public intervention	Local perception of mobility limitations and environmental quality
Approach	Top-down identification and definition of problems, and articulation of solutions	Bottom-up detection of local needs and development of solutions
Coordination	Multilevel administrative coordination	Self-organised, (occasionally) institutionally-supported activities
Aims	Regional level green infrastructure developed at strategic level	Local level green infrastructure elements adapted to locally perceived needs
Means	Institutional implementation, top-down analysis with support of information retrieved from public participation processes	Local organisations driving self-organised efforts, local networks between organisations, active participation and proposals to institutions
Identified barriers	Institutional lack of coordination; lack of sectoral coherence; short-term visions; sensibility to administrative changes	Apathy in some community institutions; divergence of visions; reactive approach to management in local administrations
Identified needs	Institutional change at different levels of governance; horizontal and vertical administrative coordination; multilevel/multiscale learning	Improved communication among some organisations and institutions; improved collaboration with local institutions; improved institutional visibility

requirements. Different potentials and constraints can be interpreted through the levers as governance mechanisms applied at regional and local scales through their related leverage points; for instance:

- Incentives and capacity building for environmental responsibility are already a central element in the educational and other initiatives analysed at the local level. Focusing on improvement of quality of life in the neighbourhood and the removal of waste serve as ways to comprehend the consequences of consumption dynamics. This allows for the potential to use GI to aim directly towards envisioning a good quality of life while lowering total consumption and waste.
- Cross-sectoral cooperation, avoiding segmented decision-making and promoting integration across jurisdictions, is one of the main needs identified in the process, and would include institutional change at the regional level, collaboration among local-level institutions, and improved communication, coordination and integration between levels. Current deficiencies in this lever are translated into ill-implemented regional strategies and local initiatives that lack reach, thus hindering the transformative potential at all scales.

- Pre-emptive action to avoid, mitigate and remedy the deterioration of nature can be seen as one of the motivations of GI implementation, motivating the development of strategies at different levels. Activation by unleashing values and action, reducing inequalities and practicing justice and inclusion in conservation as key points may boost the capacity of GI to achieve the interlinked goals of the nexus.
- Decision-making in the context of resilience and uncertainty is nowadays a big challenge at any of the studied levels. The identified barriers, related to the approach, coordination and means of strategic planning, reveal deficiencies in its adaptive character, thus being hardly capable to cope with uncertainty and complexity. Inclusion of new perspectives for strategic planning in innovative governance schemes could provide adaptive decision-making articulated through social innovation, adaptive management and similar points of action.
- Environmental law and implementation will be one of the bases for the application of GI at the regional level. At the local level, it is perceived that the enforcement of existing environmental law would overcome many of the current barriers and constraints for the improvement of environment through community-designed GI elements. Well-planned implementation of environmental law would help to unleash the latent values of responsibility and social norms for sustainability leading to action.

8.5   Conclusions

Transformations taking place at the local level are guiding a transition in the studied SEPLS from the traditional agricultural system to the current one, where the local inhabitants are developing a new relationship with their environment based on the continuous development of new visions. In this sense, bottom-up implementation of green infrastructure elements emerges from needs identified by the community regarding their wellbeing and conducted through a Neighbourhood Association. Specific actions are centred in the reclamation of riparian and forest ecosystems as multifunctional greenways. Clean-up of dumped trash, recovery of heritage sites and educational activities are developed to encourage holistic views for the transformation towards a circular economy at the local level, including employment opportunities. Key components are the involvement of the community, the development of a network of organisations, and collective leadership. Institutional support by the local and regional government is perceived, though, as essential, and its absence may cause limitations to the transformation potential of the association.

The efforts developed at the regional level addressing transformative change are mainly statutory, top-down in character, with the transposition of directives coming from upper levels being an important motivation for green infrastructure implementation. This view is complemented by public participation to integrate bottom-up perspectives and visions. As such, it could be a reference point for identifying and developing learning loops. As implementation at the regional level is associated to spatial planning, there is a strong dependence on institutional and governance structures. In this sense, barriers imposed by the lack of communication between administrations, deficits in sectorial coherence, short-term views and sensibility to changes in administration schemes evidence the need for institutional change. Such barriers may also affect initiatives developed at the local level, depriving them of institutional support. Also, they may preclude essential tasks such as assessment of the strategic interest of bottom-up initiatives, sectoral coordination, regulation enforcement, or valuation of financial opportunities. Additionally, to face the challenges involved in the transformative change, an essential role of the new governance schemes would be the identification of the priority, public and general interest of bottom-up initiatives aligned with sustainability goals to warrant their effective implementation.

As a concluding remark, the differences identified between the analysed levels (local and regional) and perspectives (community and statutory) highlight the need for deliberative, conversational approaches integrated in innovative governance schemes as a prerequisite for transformative change. Comprehension and activation of multi-actor governance 'levers' and 'leverage points' will be essential in this process.

References

Botequilha-Leitão, A., & Díaz-Varela, E. R. (2020). Performance based planning of complex urban social-ecological systems: the quest for sustainability through the promotion of resilience. *Sustainable Cities and Society, 56*, 102089.

Bouhier, A. (2001). *Galicia. Ensaio xeográfico de análise e interpretación dun vello complexo agrario*. Santiago de Compostela: Consellería de Agricultura, Gandería e Política Agroalimentaria (Xunta de Galicia).

Calvo-Iglesias, M. S., Crecente-Maseda, R., & Fra-Paleo, U. (2006). Exploring farmer's knowledge as a source of information on past and present cultural landscapes: A case study from NW Spain. *Landscape and Urban Planning, 78*, 334–343.

Calvo-Iglesias, M. S., Fra-Paleo, U., & Diaz-Varela, R. A. (2009). Changes in farming system and population as drivers of land cover and landscape dynamics: The case of enclosed and semi-openfield systems in Northern Galicia (Spain). *Landscape and Urban Planning, 90*, 168–177.

Centro Nacional de Información Geográfica. (2019). *Centro de Descargas*, Madrid, viewed 22 July 2020, http://centrodedescargas.cnig.es/CentroDescargas/index.jsp

Chan, K. M. A., Agard, J., & Liu, J. (2019). Chapter 5. Pathways towards a sustainable future. In E. S. Brondizio, J. Settele, S. Díaz, & H. T. Ngo (Eds.), *Global assessment report of the Intergovernmental science-policy platform on biodiversity and ecosystem services*. Bonn: IPBES secretariat.

Cooney, N. (2010). *Change of heart: What psychology can teach us about spreading social change*. New York: Lantern Books.

Xunta de Galicia (2019). *Información Xeográfica de Galicia*, Santiago de Compostela, viewed 22 July 2019, http://mapas.xunta.gal/portada

Xunta de Galicia (2017). *Convenio de colaboración entre a Consellería de Medio Ambiente e Ordenación do Territorio a través da Dirección Xeral de Patrimonio Natural e do Instituto de Estudos do Territorio, a Universidade de Santiago de Compostela (USC) e a Universidade da Coruña (UDC) para o deseño da estratexia de infraestrutura verde de Galicia cofinanciado nun*

80% polo Fondo Europeo de Desenvolvemento Rexional no marco do Programa Operativo FEDER Galicia 2014–2020. Galicia: Galician Regional Government (Xunta de Galicia)

Decree 42/2019 – Decreto 42/2019 de 28 de marzo, por el que se establece la estructura orgánica de la Consellería de Medio Ambiente, Territorio y Vivienda (Galicia, Spain).

DG Environment. (2012). *The multifunctionality of green infrastructure, Science for Environment Policy, DG Environment News Alert Service, In-Depth Report, March 2012*. Brussels: European Commission.

Díaz, S., Settele, J., Brondízio, E. S., Ngo, H. T., Agard, J., Arneth, A., et al. (2019). Pervasive human-driven decline of life on Earth points to the need for transformative change. *Science, 366* (6471), eaax3100.

Díaz-Varela, E., Ferreira-Golpe, M. A., García-Arias, A. I., Pérez-Fra, M., López-Iglesias, E., & Rodriguez-Morales, B. (2018). Capítulo 10: Estratexias para o aproveitamento das potencialidades da infraestrutura verde para o desenvolvemento socioeconómico. In *Estratexia de Infraestrutura Verde de Galicia (Unpublished Draft)*. Xunta de Galicia: Instituto de Estudos e Desenvolvemento de Galicia (IDEGA) – Universidade de Santiago de Compostela (USC) – Instituto de Estudos do Territorio (IET) – Consellería de Medio Ambiente Territorio e Vivenda (CMATV).

Diego Fuentes, A. (2019). *Programa transversal sobre educación ambiental, patrimonio e participación cidadá "Regatos e muiños.: paraísos baixo o lixo"*. Kitchener: Neighbourhood Association of Chapela.

Dominguez-Garcia, D., Swagemakers, P., & Simon, X. (2015). Sustainable management of green space in the city-region of Vigo, Galicia (Spain). In *Second international conference on agriculture in an urbanizing society reconnecting agriculture and food chains to societal needs*, 14–17 September, Rome.

European Commission. (2011). *Communication from the commission to the European Parliament, the Council, the European Economic and Social Committee and the Committee of the Regions. Our life insurance, our natural capital: an EU biodiversity strategy to 2020, COM (2011) 244 Final*. Brussels: European Commission.

European Commission. (2013a). *Communication from the commission to the European Parliament, the Council, the European Economic and Social Committee and the Committee of the Regions. Green Infrastructure (GI) - Enhancing Europe's Natural Capital, COM (2013) 249 Final*. Brussels: European Commission.

European Commission. (2013b). *Building a green infrastructure for Europe*. Brussels: European Commission.

European Environment Agency. (2017). *Biogeographical regions in Europe*, Copenhagen, viewed 2 November 2020, https://www.eea.europa.eu/data-and-maps/figures/biogeographical-regions-in-europe-2

Fernandez-Villar, G. (2019). *Estudo das infraestruturas verdes da parroquia de chapela e colindantes e proposta alternativa ao modelo de aproveitamento agro-forestal actual, Final Degree Project, Agricultural and Agri-Food Engineering*. Santiago de Compostela: University of Santiago de Compostela.

Grădinaru, S. R., & Hersperger, A. M. (2019). Green infrastructure in strategic spatial plans: Evidence from European urban regions. *Urban Forestry & Urban Greening, 40*, 17–28.

Granberg, M., Bosomworth, K., Moloney, S., Kristianssen, A.-C., & Fünfgeld, H. (2019). Can regional-scale governance and planning support transformative adaptation? A study of two places. *Sustainability, 11*(24), 6978.

INE. (2019). *Instituto Nacional de Estadística (Spanish Statistical Office) – Nomenclátor, Nomenclátor: Población del Padrón Continuo por unidad poblacional*, viewed 11 March 2020, http://bit.ly/2WToH7L

Infraestructura Verde de Galicia. (2019). *Infraestructura Verde de Galicia*, Santiago de Compostela, viewed 9 March 2020, http://infraestruturaverdegalicia.gal/

IPBES. (2019a). *IPBES rolling work programme up to 2030*, IPBES, viewed 11 March 2020, https://ipbes.net/work-programme

IPBES. (2019b). Summary for policymakers of the global assessment report on biodiversity and ecosystem services of the Intergovernmental Science-Policy Platform on Biodiversity and Ecosystem Services. In S. Díaz, J. Settele, E. S. Brondízio, H. T. Ngo, M. Guèze, J. Agard, A. Arneth, P. Balvanera, K. A. Brauman, S. H. M. Butchart, K. M. A. Chan, L. A. Garibaldi, K. Ichii, J. Liu, S. M. Subramanian, G. F. Midgley, P. Miloslavich, Z. Molnár, D. Obura, A. Pfaff, S. Polasky, A. Purvis, J. Razzaque, B. Reyers, R. R. Chowdhury, Y. J. Shin, I. J. Visseren-Hamakers, K. J. Willis, & C. N. Zayas (Eds.), *IPBES Secretariat*. Bonn: IPBES.

Lennon, M., & Scott, M. (2014). Delivering ecosystems services via spatial planning: reviewing the possibilities and implications of a green infrastructure approach. *Town Planning Review, 85*(5), 563–587.

Lois-González, R. C., & Aldrey-Vázquez, J. A. (2010). El problemático recorrido de la ordenación del territorio en Galicia. *Cuadernos Geográficos, 47*(2), 583–610.

McAlpine, C. A., Seabrook, L. M., Ryan, J. G., Feeney, B. J., Ripple, W. J., Ehrlich, A. H., et al. (2015). Transformational change: creating a safe operating space for humanity. *Ecology and Society, 20*(1), 56.

Morán-Ordóñez, A., Suárez-Seoane, S., Calvo, L., & de Luis, E. (2011). Using predictive models as a spatially explicit support tool for managing cultural landscapes. *Applied Geography, 31*, 839–848.

Neighbourhood Association of Chapela. (2019). *Propostas para o PMUS en Chapela*. Kitchener: Neighbourhood Association of Chapela.

Ronchi, S., Arcidiacono, A., & Pogliani, L. (2020). Integrating green infrastructure into spatial planning regulations to improve the performance of urban ecosystems. Insights from an Italian case study. *Sustainable Cities and Society, 53*, 101907.

SIOSE. (2011). *Sistema de Información sobre Ocupación del Suelo de España (Spanish Land Use Information System)*, Ministerio de Transportes, Movilidad y Agenda Urbana, viewed 11 March 2020, https://www.siose.es/

Small, D. A., Loewenstein, G., & Slovic, P. (2007). Sympathy and callousness: The impact of deliberative thought on donations to identifiable and statistical victims. *Organizational Behavior and Human Decision Processes, 102*(2), 143–153.

Souto-González, X. M. (1993). A política territorial en Galicia: entre a expansión urbana e a percepción rural. O caso do espacio periurbano de Vigo. *Minius, II-III*, 199–222.

Swagemakers, P., & Dominguez-García, D. (2015). How to move on? Collective action and environmental protection in the city-region of Vigo, Spain. In *International Conference Meanings of the Rural – between social representations, consumptions and rural development strategies* (pp. 1–5). Aveiro: University of Aveiro.

Tubío-Sánchez, J. M., & Crecente-Maseda, R. (2016). Forcing and avoiding change. Exploring change and continuity in local land-use planning in Galicia (Northwest of Spain) and The Netherlands. *Land Use Policy, 50*, 74–82.

Valladares, F., Gil, P., & Forner, A. (Eds.). (2017). *Bases científico-técnicas para la Estrategia estatal de infraestructura verde y de la conectividad y restauración ecológicas*. Madrid: Ministerio de Agricultura y Pesca, Alimentación y Medio Ambiente.

Chapter 9
Water with Integrated Local Delivery (WILD) for Transformative Change in Socio-Ecological Management

Jasmine E. Black, Chris Short, and Jenny Phelps

Abstract An innovative approach towards transformative change through multi-stakeholder participation for socio-ecological practices—Integrated Local Delivery (ILD)—has been used to restore the water quality and biodiversity across a catchment in the Cotswolds, South West England. This was triggered by the need to improve the Ecological Status of water as a part of the European Union's Water Framework Directive. On a landscape scale of roughly 25,000 hectares, multi-stakeholders collaborated through a bottom-up approach to carry out environmental restoration of the catchment.

Over 3 years, an iterative learning loop of reflection and evolution created increased engagement. Twenty farmers have been empowered as 'guardians' to be key contacts between institutions and ensure the sustained environmental quality of the area. Both farmers and communities acted to reduce chemical use, protect river banks from livestock damage and clear waterways to enhance water quality and biodiversity. Local communities fed into the development of a 'Community Water Guide' which can be applied internationally for similar projects. Within the Inter-governmental Science-Policy Platform on Biodiversity and Ecosystem Services (IPBES) transformative change framework, the ILD model can also be applied by facilitators to access levers and leverage points in order to enable change.

Important take home messages from the project include having well-trained facilitators who ensure active engagement, connections and continuity over the long term. Likewise, ensuring all stakeholders feel listened to and clearly communicated with is essential to build trust and motivation.

Keywords Integrated local delivery · Transformative change · Catchment scale · Socio-ecological · Landscape

J. E. Black (✉) · C. Short
Countryside and Community Research Institute, Cheltenham, Gloucestershire, UK
e-mail: jblack2@glos.ac.uk

J. Phelps
Farming and Wildlife Advisory Group, Tetbury, Gloucestershire, UK

M. Nishi et al. (eds.), *Fostering Transformative Change for Sustainability in the Context of Socio-Ecological Production Landscapes and Seascapes (SEPLS)*,
https://doi.org/10.1007/978-981-33-6761-6_9

155

9.1 Introduction

Since the Rio Earth Summit in 1992, there has been a sequence of calls towards more sustainable management of the planet's resources. This has included the Millennium Ecosystem Assessment in 2005 highlighting the depletion of resources by human activity, the work of the Intergovernmental Panel on Climate Change (IPCC), namely reports on the effects of 1.5 °C warming, and the recent 2019 Intergovernmental Science-Policy Platform on Biodiversity and Ecosystem Services (IPBES) Global Assessment Report on the need for transformational change (MEA 2005; de Coninck et al. 2018; IPBES 2019). Action has been encouraged at global and national levels to enact frameworks, such as the Convention on Biological Diversity (CBD) and IPBES' transformational change framework (Tengö et al. 2017; Kozar et al. 2019). With the next IPBES assessment from 2021 onwards, a key question is why has transformative change not yet happened despite these frameworks being put in place.

Transformative change is a challenging concept for the top-down institutional and governance models of recent times; to be effective, it requires working across disciplines and institutional levels as well as the wider landscape. Inherently this is more complex than a single resource and a single discipline and institutional level undertaking management, therefore a 'collective framing' and an 'integrated approach' is needed (Thiel et al. 2015; Gualini 2018; IPBES 2019). Governance therefore also needs to be adaptive, allowing for communities to take ownership of actions on the ground, opposing top-down governance (Short 2015). Through this process, institutions learn that local communities are able to undertake effective resource management alongside other actors, which then influences these institutions and policies (Carlsson and Berkes 2005; Ostrom 2005). Importantly, social capital, trust, shared values and legitimacy are needed at local scales in order to make action effective (McAreavey 2006). Across disciplinary and institutional scales, these assets are also needed for good integration, collaboration and adaptation for landscape-scale resource management (Blackstock et al. 2014). Through working effectively together in this way, the range of societies and ecologies in a landscape, as well as the associated knowledge and governance mechanisms, can be transparently addressed to work out trade-offs in tangible and non-tangible benefits (IPBES 2019). This is an iterative learning loop, in which reflective practice and adaptation strengthen and progress socio-ecological systems (Alexander 2006; IPBES 2019).

The project assessed in this chapter used the Integrated Local Delivery (ILD) model, driven by the European Union's (EU) Water Framework Directive (WFD), that called for all water courses to have 'Good Ecological Status' (GES) within a certain timeframe, which for the UK had been set at 2027. The WFD also encourages EU member states to take an approach which incorporates all stakeholder levels, from communities to policy as well as all process levels, from planning to implementation (Healey 1998). These inclusive co-developed methods of working holistically are progressing on national-global levels (Short 2015; Thiel et al. 2015). The ILD model method spans disciplines, institutional levels and landscape scales. It also allows for a critical assessment of its effectiveness, through 'action research'. ILD has been applied to a range of projects and is easily transferrable to any community.

Here we set out the ILD approach and show its effectiveness within the Water with Integrated Local Delivery (WILD) project. We also assess how the project utilises levers and leverage points across a diversity of stakeholders and ecological systems according to the IPBES framework (IPBES 2019).

9.2   The WILD Project

The WILD project was initiated in South West England within the Upper Thames river catchment area through a partnership between the Farming and Wildlife Advisory Group South West (FWAG SW), the Gloucestershire Rural Community Council (GRCC), the Cotswold Water Park Trust and the Countryside and Community Research Institute (CCRI), which is part of the University of Gloucestershire. The project ran through two stages, initially from 2013–2016 and then from 2016–2018. The project facilitator (FWAG) carried the framework over from a previous project to WILD. The facilitator involved has existing links with farming communities in the area and a history of working with them towards more environmentally sustainable practices. As a number of waterbodies in the case study area had been assessed as either moderate, poor or bad by the Environment Agency (EA) of the UK government, under the EU's WFD, the project's main objective was to improve the conditions and bring them to GES through an integrated approach. Its aims were:

- To deliver GES through ILD and direct actions, informed by catchment reconnaissance trips and advisory visits, in waterbodies in the WILD project area in line with the Upper Thames Catchment Management Plan.
- To create a framework to address other negative drivers impacting on water quality in the medium (2021) and long term (2027) to achieve GES in all surface and ground waterbodies as of EU Directives.
- To embed and enable local delivery so that the protection of water quality becomes self-sustaining, through awareness and collective responsibility of the locales.
- To integrate and deliver the relevant objectives of partners using ILD.
- To assess the project's effectiveness to inform future funding programmes and decision making of the multi-stakeholders involved.

The case study site covers a diverse landscape of small urban settlements and rural areas of agricultural land, meadows, woodland, wetlands and waterbodies. This diversity has necessitated the incorporation of multiple stakeholders spanning a wide range of institutional and structural levels. It has therefore also been vital for these stakeholders to develop strong relationships, empathy and understanding of values and goals.

9.2.1 Case Study Site

The WILD project site covers about 26,000 ha within the river basin catchment area of the River Thames (Fig. 9.1). It lies within the Cotswolds and Upper Thames Clay Vales 'Natural Character Area', designated to highlight similar national characteristics for landscape-scale management (Natural England 2015). Further, internationally important lowland meadows, limestone grassland and wetland habitats have been designated (as of the England Biodiversity Action Plan, Post-2010 Biodiversity Framework UK, under EU Biodiversity Targets and CBD Aichi Targets). The geology is predominantly limestone, providing significant groundwater aquifers, although clay is also present. The majority of the area is rural which includes agricultural land (70%) and woodland (<10%). Of the agricultural land, arable farming accounts for 43% and grassland 29%, whilst urban areas make up 15%. Towns and villages, which historically form administrative parish districts, are present and are impacted through floods and pollution. The agricultural land and industry create pollution, run off, discharges and eutrophication giving the area a Nitrate Vulnerable Zone (NVZ) status since 2002. Limestone allows pollution to flow into groundwater, whereas clay increases run-off through impermeability. The Catchment Sensitive Farming programme was set up by the government to allow farmers to voluntarily reduce pollution, which then developed a Catchment Based Approach which has been built upon by the WILD project.

9.3 Methods

9.3.1 The ILD Approach

The WILD project used the Integrated Local Delivery (ILD) model led by FWAG to bring together multi-stakeholders starting from the community level and working upwards. There are six key steps to the ILD, which are aimed at the facilitating organisation or individual, described in Table 9.2 below. Vital background scoping of environmental and cultural assets, issues, values and capacities is undertaken using the diagram in Fig. 9.3, which shows movement from identifying the inner local issues and assets, out to the regional, national and international (A–D). In relation to these issues, identification of the individual or institutions responsible for the delivery of frameworks and policy related to the identified issues then takes place, moving in reverse from the international to local scales (E–H) (Short et al. 2010).

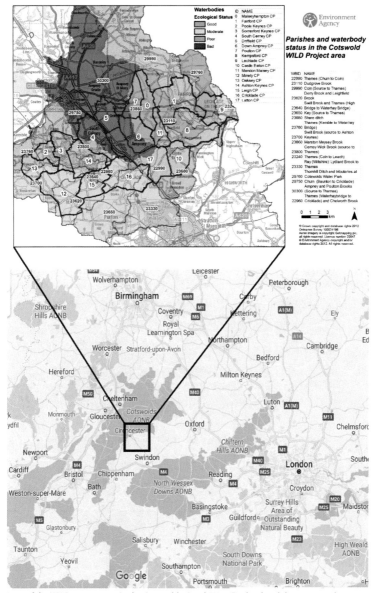

Map of the WILD project area in the Cotswolds, South West England and the associated ecological status of the waterbodies.

Fig. 9.1 Location of the WILD project area and the corresponding map of waterbodies included in the project, as classified by the Environment Agency (EA) prior to the work being carried out (figure adapted from Google Maps, 2020 and the Environment Agency, UK Government, and Ordinance Survey 2012)

Fig. 9.2 Livestock grazing grassland at a group visit in the WILD catchment area (photo: Jenny Phelps)

Table 9.1 Basic information of the case study area

Country	UK
Province	Gloucestershire
District	Cotswolds
Municipality	n.a.
Size of geographical area (hectare)	22,692
Number of indirect beneficiaries	192,441
Dominant ethnicity(ies), if appropriate	British
Size of case study/project area (hectare)	22,692
Number of direct beneficiaries	1000
Dominant ethnicity in the project area	British
Geographic coordinates (latitude, longitude)	51° 43′ 6.60″ N; 1° 58′ 5.52″ W

Table 9.2 Integrated local delivery model process (adapted from Short et al. 2010)

1. Once invited, begin initial scoping to determine the area, its assets, key individuals and strategic frameworks involved	• Before visiting the area collect background strategic, historical and cultural information to scope the area's assets and core issues • Start with an open mind and determine the administrative area that includes all legal stakeholders and local interests (e.g. parish or ward) • Gather many views in order to gain a comprehensive understanding of both assets and uses of the area with contacts for each • Aim to try and understand local custom and tradition which influences the way in which the community works and how various decisions are made at the local level. Value this information
2. Map the management tasks and verify these in an inclusive and open format	• Bring the findings from step 1 to the community so that local knowledge and data can contribute to and strengthen the information you have found • Confirm the spatial area with the community and government agencies and the key assets, issues and challenges to be resolved • Outline the opportunities so the local stakeholders and community clearly understand what tasks and challenges could be achieved together. Be enthusiastic and realistic
3. Develop a management group around key local and statutory stakeholders	• Disseminate proposals arising from step 2 through local meetings, informal discussions and guided walks with local, regional/national stakeholders • Develop a transparent and inclusive local management structure that sits within the existing administrative framework • Confirm arrangements with regional and national statutory bodies and other agencies and ensure support for management proposals and acknowledgment of the importance and benefit of local knowledge
4. Encourage linkages and opportunities for local contribution and adoption of responsibilities	• Identify strategic priorities from step 3 that might be delivered by the local management group and associated funding streams and opportunities • Enable local responsibility through partnership working with appropriate statutory agencies alongside an associated funding plan • Ensure opportunities for local ownership with key responsibilities led by local group alongside support of statutory agencies
5. Establish capacity and role of the local management group; identify and prioritise tasks	• Once step 4 is agreed, having identified a management structure and responsibilities, support the local group to take the lead

(continued)

Table 9.2 (continued)

	• Identify features and tasks that can be used to develop the capacity (both skills and commitment) of local and statutory stakeholders • Establish the role of the local group so it fulfils requirements of public bodies/associated funding responsibilities and is recognised as a subcommittee of the agreed administrative unit (e.g. parish council)
6. Implement proposals and embed management group and support	• After step 5, it is for the management group to agree which actions to prioritise through funding and overall implementation process • Determine the most appropriate local government link to embed the group within a transparent and accountable structure • Enable group members to offer their contribution and resources, allocation of specific tasks and training opportunities for volunteering • Support early implementation and discuss the process with local group and statutory agencies to ensure group is working effectively • Agree with the group further points for internal review and ensure statutory agency availability to discuss issues on-site and remotely • Check for equity, balance and inclusion in local group

9.3.2 ILD Applicability to IPBES Transformative Governance Framework

Within the IPBES framework for transformative change, 'levers' and 'leverage points' highlight where and what interventions can be put in place to achieve mutually reinforcing, larger-scale changes towards more sustainable governance at multiple structural levels. The ILD approach allows these levers and leverage points to be identified in order for environmental management practices to be applicable to this global governance framework. A matching of these levers and leverage points with the WILD project has been undertaken.

9.3.3 ILD Evaluation: Survey and Interviews

The first stage of the WILD project (2013–2016) has been evaluated through the use of surveys and interviews (to gain a deeper understanding) across all stakeholder groups. In both year one (2013–2014) and year 3 (2015–2016) surveys and

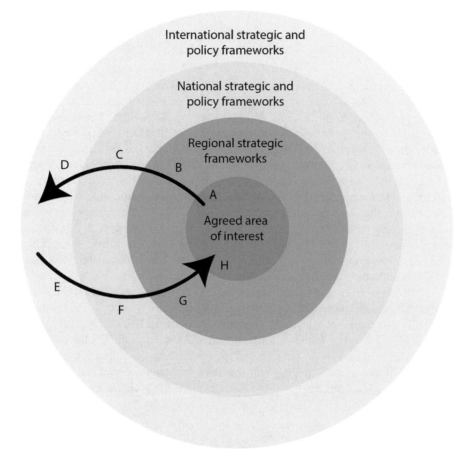

Fig. 9.3 Scoping phase within integrated local delivery (adapted from Short et al. 2010)

interviews were held with the community groups, farmers and partners involved. The results from the final year have been used for this paper, with the addition of any important notes on the initial 2013–2014 survey where relevant. Action research was undertaken through researchers attending group meetings and volunteer days to understand participants' views. The evaluation intended to capture the process of interaction, level of equality in governance and outcomes between the various organisations, interest groups and communities through the wider and more integrated approach to water management. The evaluation framework also accounted for the ecological aspects (GES) of the project, undertaken by the EA.

9.4 Results

9.4.1 Overall Project Results

A total of 298 farm visits over 3 years were undertaken, with advice provided for 118 farms covering 22,692 ha of land. Of the 19 parishes in the catchment area, 16 engaged, as well as three towns, incorporating WILD into their Parish Planning. Further, 20 'farmer guardians' were appointed over 10,000 ha who acted as key contacts liaising between the farming community and other stakeholders such as the EA and independent company Thames Water. Part of the project included reducing the amount of regulation from the EA and replacing this with a more collaborative approach, which empowered farmers to take their own action alongside support from other stakeholders. Local communities and members of the agencies involved contributed the equivalent of 216,000 GBP of volunteer time over 3 years. Communities discussed over 1500 issues regarding water flow, which were mapped and digitised. Management measures within the river were undertaken on 5580 m of river, tree pollarding and shade-reduction along 8555 m of river (Fig. 9.5), fencing along 5066 m of river, and five livestock drinking bays were installed (Fig. 9.4). "A Community Guide to Your Water Environment" was also produced and circulated nationally as a toolkit to Integrated Planning alongside the ILD.

Under the EU Water framework directive, the EA surveys the Ecological Status (ES—high, good, moderate, poor or bad) of the waterbodies every 3 years. ES is measured through the quality of multiple indicators including biological (e.g. phytoplankton, macroalgae, invertebrates, and fish), chemical and physico-

Fig. 9.4 Soil erosion mitigation measure (drinking bay) for cattle when accessing the river (photo: Jenny Phelps)

Fig. 9.5 Better accessibility to waterbodies via the construction of a boardwalk for local communities (photo: Jenny Phelps)

Fig. 9.6 Local community volunteers clearing riverside branches to improve water quality (photo: Jenny Phelps)

Fig. 9.7 Herbal leys sown by WILD farmers to enhance soil health (photo: Jenny Phelps)

chemical, water (synthetic and non-synthetic pollutants) and hydromorphological. Following the first phase of the project, an improvement or lack of deterioration in the majority of waterbodies was measured. Both waterbodies initially deemed 'bad' improved to 'moderate' and two out of four initially deemed 'poor' improved to 'moderate', the other two remaining 'poor'. Of those deemed 'moderate' initially, four out of six have remained so, whilst two have deteriorated to 'poor' or 'bad'. It is important to note that the measurements taken to determine waterbody ES changed over the duration of the project, and therefore may have had an impact on these results. Depending upon when improvements to waterbodies were made, there may also be a lag in improvements (or otherwise) being seen, which would thus show in the next cycle of EA assessments. It is evident that there is still work to be done to improve and maintain some of the waterbodies up to GES through the ILD approach. Importantly, the evident deterioration of some waterbodies is due to poor agricultural practices such as livestock management, drainage and physical barriers to ecology.

9.4.2 Survey and Interview Results

9.4.2.1 WILD Partners

The partners surveyed included 15 respondents across eight organisations, of which a total of seven (47%) completed the final survey in 2015–2016. Most partner respondents felt that they had either a medium or high involvement with the WILD project and had a good understanding of the multiple objectives and aims of the project, as expected.

The respondent partners felt that the project's main achievement, aside from progression towards GES, was strong social cohesion amongst all stakeholders involved. This included developing local networks, accountability and responsibility, tools and skills to improve resilience. The ILD management approach was thought to have been inclusive and information effectively shared amongst the stakeholders. In the first year, stakeholders voiced the need for better collaboration and dialogue across the project, however this had improved by the second and third years, with six out of seven partners expressing that, "*the main project group collaborates and organises joint activities, fully integrating multiple areas of interest*". An important point regarding the reduction of farmer regulation to reduce the stress of breaching rules and nurture more environmentally respectful attitudes was voiced by one partner: "*. . .there is no question that an approach which provides friendly advice is preferable to one which relies on enforcement, but both are required.*" Regarding community involvement, partners reflected that despite it taking some time to work within local communities, they were able to see the benefits as the project progressed. One respondent remarked that "*. . .it is the individuals involved in the delivery, as much as the delivery model itself, which is crucial to a partnership project's success.*" Key findings from the partners were that local communities matter and communication is vital. Further, integration of shared interests was thought to strengthen funding possibilities and promote project aim delivery.

9.4.2.2   Local Authorities and Councils

There was a total of six responses out of 32 potential respondents across five local authorities (19% response rate) to the final year survey (2015–2016). Most respondents (four out of six) stated high involvement with the project, with just one stating low involvement.

The respondent local authorities felt that the WILD project broadly focused on improving community relations, improving water quality and reducing flood risk (creating better ecosystem understanding). The collaborative approach was thought of as beneficial from the perspective of most respondents. However, the need for a greater amount of funding and a longer timeframe was voiced, as building relationships with communities takes time but, crucially, makes the project effective. The local authority respondents also had mixed views on whether the project had incorporated their views and interests; they did not always feel a strong relationship with the project. Despite this, they did think that community groups had equal opportunities to be involved. They further thought that good communication needed to be maintained through more feedback, and that the project lacked the relevant technical expertise for the ecological management of waterbodies. Regardless, the opinion was that the ILD should be used elsewhere across the country. It was thought to add value to the existing 'conventional' regulatory, institutionally-led (top-down) approach with fewer stakeholders. ILD's implementation needs careful consideration of such previous approaches and resulting attitudes and relationships. One respondent commented, "*ILD is working with people and that will always be preferred to enforcement through improved local involvement*".

9.4.2.3 Town Councils and Communities

A total of ten out of 19 parish and town councils responded (53%) to the final 2015–2016 survey. Only two out of ten respondents stated their involvement as low, two out of ten as high, and the remainder as medium.

For local communities, the main aims of the project were to improve knowledge and understanding of the water situation in the catchment and improve its quality and resilience to flooding. Most respondents felt that this had been achieved, whilst others perceived it as a continuous process. As the first stage of the project was short, there was not a perceived difference in the water quality by local communities. Despite this, they felt engaged and empowered to enact environmental management as WILD had provided crucial knowledge and support through key stakeholders who imparted this: "*We are much better informed about surface water flow in the area, and this should enable us to decrease flood risk*". Community members now know who to contact if support is needed. Their awareness of the water environment has also increased, whilst activities such as mapping waterbodies and courses as well as clearing ditches were very valuable skills to learn. Key to this was the ability to create a work and maintenance plan, as well as involvement in neighbourhood plans for new building developments. Community members also felt able to undertake future work regarding the water environment, however concerns were expressed over funding and capacity. They felt that engagement with farmers had been both "*constructive and effective*". The facilitation team was considered very valuable but had big time pressures which have the potential to affect future work through reducing enthusiasm and motivation.

9.4.2.4 Farmers and Land Owners

Across 100 farms there were five responses out of 22 potential respondents (23%) to the final 2015–2016 survey. A further ten farmers and land managers were interviewed face-to-face and views sought from a number of farmer-to-farmer meetings.

Over the course of the first 3 years, farmers and land owners felt that WILD had increased their knowledge and awareness of ecological management: "*WILD has made us more aware of water movement and water management, so it has been positive*". Further, it has improved their awareness on sources of advice, volunteer labour and potential benefits to the community. Over the 3 years, farmers' and land owners' relationships to WILD improved, as more became involved and challenges decreased. Several respondents noted that the project was aiding their goals for national agri-environment schemes, which aim to encourage environmental land stewardship. It was therefore able to assimilate well with existing governance and policy mechanisms. Respondents voiced concerns over time costs, however these were mitigated by volunteers in undertaking management actions. The respondents acknowledged the beneficial capacity of volunteers. Despite this, there was some feeling of lower transparency, as volunteers *"aren't specialists"*, with a clash in priorities: *"...we need to monitor the environment, but emphasis is on increasing biodiversity, which conflicts with my view as an arable farmer"*. However, WILD was seen as preferable to conventional national schemes due to its collaborative approach with reduced regulation, more flexibility and less administration. Farmers and land owners underwent behaviour change, embracing new techniques and approaches: *"If there were an identical project to come along now, and knowing all the input I would have to do, I would still do it"*.

9.5 Discussion

9.5.1 Change, Challenges and Opportunities of WILD

The main achievement of the WILD project has been to include all of the local and national interests in the discussion on priority water environment challenges, and to implement some measures within a 3-year period. There was no local resistance to the project, potentially due to the open and inclusive approach taken. Prior to this project, government schemes had fragmented management and very little multi-stakeholder discussion. This key difference highlights the importance of a qualified independent facilitator to inform, create networks, enable and crucially embed the process across multiple stakeholders and institutional levels. Most stakeholders involved felt engaged by the project and that its aims had either been achieved or were in progress. There was also an acknowledgement that this project involves long-term thinking and action, which needs time and support to build relationships, actions and reflections in order to evolve. In embedding the ILD approach across several localities, the facilitators have achieved a local self-governing system which is creating behaviour change in farmers, land owners, local community members as well as at institutional levels in local authorities, government bodies and organisations (Alexander 2006; IPBES 2019). However, there is still need from some stakeholders for facilitation, time and funding in order to progress the work which has begun. For those farmers and land owners who felt the process was less transparent than other schemes run by the government, there may still be progress to be made in communications between them and local communities. It will also be important for long-term monitoring of waterbodies and any ecological improvement to be presented and discussed by the WILD stakeholders in future years. Whilst funding for the following 3 years of the project (2016–2018) had been secured through a combination of the EA and private investment (Thames Water), beyond this funding, continuation is uncertain, especially in light of the UK separation from the EU. This funding has not extended to qualitatively evaluating the second stage of the project, therefore unfortunately the stakeholders' perceptions and progress are yet unknown. The future uncertainty of the project needs to be addressed and a way found for multi-stakeholders to continue using an ILD approach to local and regional environmental issues.

9.5.2 ILD within IPBES Transformative Governance Framework

Table 9.3 shows how ILD fits within the IPBES governance and policy transformative change framework, identifying the levers and leverage points that were utilised during the WILD project. In fitting the ILD model to the IPBES framework, it is apparent that facilitation is a vital, underpinning factor to the cohesion of multi-stakeholders and the success of the project. In this context, the ILD and its facilitators build key 'levers', including capacity, cross-sectoral cooperation, decision-making processes and the ability for pre-emptive action in the future through networking and organisation of the project team. They also allow for awareness around environmental laws and making sure these are implemented by connecting communities to relevant institutions and their knowledge. The results above further show that the ILD model engages at 'leverage points' as wide stakeholder inclusion, striving for equality, respecting diverse values and building capacity and awareness around environmental issues is inherent to its process (IPBES 2019). Despite this, the results from participating stakeholders do show that this is a process that needs development with some groups, such as local authorities, needing greater engagement, and crucially facilitators needing more capacity to work with the project. Parallel to this is the need for funding which can create and strengthen capacities. Further, WILD has been able to undertake the IBPES iterative learning loop, as also demonstrated in Pahl-Wostl et al. (2008) study on social learning and culture in water management, through successive years of implementation, discussions, reflection and flexibility to evolve. Through its inherent process and iterative learning, WILD has begun the process of transformative change. It is apparent that WILD has directly incorporated global-level Sustainable Development Goals (SDGs) 2, 6, 9, 11 and 13 and Convention on Biodiversity targets 1, 3, 4, 7–9, 11, 14, 15, 18 and 19, which may help those implementing it to gain further international environmental recognition in conjunction with IPBES.

9.6 Conclusion

The WILD project has shown that the ILD model can be used to contribute to both national and international socio-ecological goals through an inclusive process. It has set in motion transformative change locally, but also further afield through the involvement of institutional partners. It deserves both more attention and scrutiny

Table 9.3 WILD project levers and leverage points (summarised by J.E. Black) within the IPBES framework (IPBES 2019)

IPBES levers	WILD example	IPBES leverage points	WILD example
Incentives and capacity building	Facilitated process, key liaison figures identified, capacity for horizontal collaboration and volunteers to enact physical management work	Embrace diverse visions of a good life	Value knowledge, priorities and needs of local and farming communities whilst incorporating other stakeholder views and ecological opportunities
Cross-sectoral cooperation	ILD model—community to policy level, with facilitators creating connections	Reduce total consumption and waste	Waste of agri-chemicals/ nutrients in farming
Pre-emptive action	ILD model could be used to prevent deterioration of good ecological status sites	Unleash values and action	Local and farming community values and knowledge used to inform action, ILD assessed via participatory action research
Decision-making in the context of resilience and uncertainty	Inclusive multi-stakeholder group decision making on issues faced in socio-ecological landscapes, e.g. flood risk	Reduce inequalities	All actors treated as equals, incorporates diverse engagement and stakeholders
Environmental law and implementation	The WILD project stemmed from millennium ecosystem assessment and EU water framework directive regulations	Practice justice and inclusion in conservation	All relevant stakeholder input valued and community empowered to lead project
		Internalise externalities and telecouplings	Socio-ecological scope of ILD; discussions including stakeholder values and ecosystem issues
		Ensure environmentally friendly technology, innovation and investment	Investment in sustainable management by water company; man-powered technology through voluntary restoration works and natural materials to prevent soil erosion (e.g. drinking bays)
		Promote education as well as knowledge generation and sharing	Meetings to share and value all actors' knowledge, development of community guide for the water environment

and should be part of the wider discussion on transformative change and alternative governance approaches. There are short-term needs in order to meet current, urgent agricultural and environmental challenges, but also longer-term issues that need to be explored through ILD, such as ensuring future motivation and discussing feedback on environmental outcomes. The emphasis moving forward should be in delivering landscape-scale change and empowering communities to take ownership of this process.

Crucially, the model requires a trained facilitator with the trust of the local and agricultural communities, or key connectors within them, as well as time and funding in order to be successful. Therefore, long-term visions and resources need to be planned into ILD projects. This needs recognition from governments, who can also incentivise businesses to help garner funding. Demonstrating contribution to international frameworks such as IPBES, SDGs and CBD targets could help ILD projects to secure funding and influence further global uptake of the model.

Acknowledgements The WILD project was funded by the Environment Agency and its undertaking has been reliant on many individual volunteers, farmers, parish and town councils' organisations and interest groups. Thanks are due to the main delivery partner, the Farming and Wildlife Advisory Group, and key delivery partners: Cotswold Water Park Trust, Gloucestershire Rural Community Council and Countryside & Community Research Institute and the National Farmers Union. Thanks finally to Thames Water who were an interested party in phase 1 of WILD, and provided funding in phase 2.

References

Alexander, E. R. (2006). Institutional design for sustainable development. *Town Planning Review, 77*(1), 2. https://doi.org/10.3828/tpr.77.1.2.

Blackstock, K. L., Waylen, K. A., Marshall, K. M., & Dunglinson, J. (2014). Hybridity of representation: Insights from River Basin management planning in Scotland. *Environment and Planning C: Government and Policy, 32*(3), 549–566. https://doi.org/10.1068/c11261.

Carlsson, L., & Berkes, F. (2005). Co-management: Concepts and methodological implications. *Journal of Environmental Management, 75*(1), 65–76. https://doi.org/10.1016/j.jenvman.2004.11.008.

de Coninck, H., Revi, A., Babiker, M., Bertoldi, P., Buckeridge, M., Cartwright, A., et al. (2018). Chapter 4: Strengthening and implementing the global response.

Gualini, E. (2018). *Planning and the intelligence of institutions: Interactive approaches to territorial policy-making between institutional design and Institution-building.* London: Routledge. https://doi.org/10.4324/9781315201726.

Healey, P. (1998). Collaborative planning in a stakeholder society. *Town Planning Review, 1998*, 1–21. https://doi.org/10.3828/tpr.69.1.h651u2327m86326p.

IPBES. (2019). *Global assessment report on biodiversity and ecosystem service, debating nature's value.* https://doi.org/10.1007/978-3-319-99244-0_2.

Kozar, R., Galang, E., Alip, A., Sedhain, J., Subramanian, S., & Saito, O. (2019). Multi-level networks for sustainability solutions: The case of the International Partnership for the Satoyama Initiative. *Current Opinion in Environmental Sustainability, 39*, 123–134. https://doi.org/10.1016/j.cosust.2019.09.002.

McAreavey, R. (2006). Getting close to the action: The micro-politics of rural development. *Sociologia Ruralis, 2006*, 85–103. https://doi.org/10.1111/j.1467-9523.2006.00407.x.

MEA. (2005). *Ecosystems and human well-being: Health synthesis*. Washington, DC: Island Press. https://doi.org/10.1016/B978-0-12-809665-9.09206-5.

Natural England. (2015). National Character Area profile: 107. Cotswolds.

Ostrom, E. (2005). *Understanding institutional diversity*. Oxford: Princeton University Press.

Pahl-Wostl, C., Tàbara, D., Bouwen, R., Craps, M., Dewulf, A., Mostert, E., et al. (2008). The importance of social learning and culture for sustainable water management. *Ecological Economics, 64*(3), 484–495. https://doi.org/10.1016/j.ecolecon.2007.08.007.

Short, C. (2015). Micro-level crafting of institutions within integrated catchment management: Early lessons of adaptive governance from a catchment-based approach case study in England. *Environmental Science and Policy, 53*, 130–138. https://doi.org/10.1016/j.envsci.2015.06.009.

Short, C., Griffiths, R., & Phelps, J. (2010). *Inspiring and enabling local communities: An integrated delivery model for localism and the environment. Report to farming and wildlife advisory group and Natural England*. Cheltenham: CCRI.

Tengö, M., Hill, R., Malmer, P., Raymond, C. M., Spierenburg, M., Danielsen, F., Elmqvist, T., & Folke, C. (2017). Weaving knowledge systems in IPBES, CBD and beyond—Lessons learned for sustainability. *Current Opinion in Environmental Sustainability, 26–27*, 17–25. https://doi.org/10.1016/j.cosust.2016.12.005.

Thiel, A., Mukhtarov, F., & Zikos, D. (2015). Crafting or designing? Science and politics for purposeful institutional change in social–ecological systems. *Environmental Science & Policy, 53*, 81–86. https://doi.org/10.1016/j.envsci.2015.07.018.

Chapter 10
Traditional Landscape Appropriation of Afro-Descendants and Collective Titling in the Colombian Pacific Region: Lessons for Transformative Change

Mauricio Quintero-Ángel, Andrés Quintero-Ángel, Diana M. Mendoza-Salazar, and Sebastian Orjuela-Salazar

Abstract The Colombian Pacific region is one of the most biodiverse areas in the world, but several anthropic pressures threaten its ecosystems and the ethnic groups who live there. Since the colonial era, the region has experienced two different key strategies of landscape appropriation: (1) diversification of activities in the landscape; and (2) specialisation focusing on a few landscape products. These two strategies fall at opposite ends of a modified continuum over time, including a range of intermediate situations that combine elements of the diversified and specialised strategies. The first strategy is characteristic of Afro-descendant communities, based on harmony with nature and favoring human well-being, while providing multiple ecosystem services and cultural or spiritual values.

In this context, this chapter reviews the relationship of Afro-descendants with their environment in the Colombian Pacific region, taking as an example the San Marcos locality. Through interviews with key informants and participant observation, we investigate the productive and extractive practices in San Marcos. Results show that the appropriation strategy combines different sources of income. This denotes a great local ecological knowledge geared to maintenance of biodiversity. Despite Law 70 (1993) stipulating Afro-descendant communities to have guaranteed autonomy and the right to collectively manage their ancestral lands, this socio-ecological production landscape is endangered due to pressures from the dominant society towards conversion to a specialised strategy. Finally, we also analyse "transformative change" in the context of governance of San Marcos. Such change

M. Quintero-Ángel · D. M. Mendoza-Salazar
Universidad del Valle, sede Palmira, Palmira, Valle del Cauca, Colombia
e-mail: mauricio.quintero@correounivalle.edu.co; diana.m.mendoza@correounivalle.edu.co

A. Quintero-Ángel (✉) · S. Orjuela-Salazar
Corporación Ambiental y Forestal del Pacífico – CORFOPAL, Cali, Valle del Cauca, Colombia
e-mail: direccioncientifica@corfopal.org; direccionejecutiva@corfopal.org

© The Author(s) 2021

175

M. Nishi et al. (eds.), *Fostering Transformative Change for Sustainability in the Context of Socio-Ecological Production Landscapes and Seascapes (SEPLS)*,
https://doi.org/10.1007/978-981-33-6761-6_10

could guide a profound transformation in conservation strategies based on a fundamental reorientation of human values.

Keywords Land use · Appropriation · Land-use transitions · Collective titling · Afro-descendants · Chocó biogeographic region

10.1 Introduction

Human interventions in the natural world correspond to what some authors have called landscape appropriation (González de Molina and Toledo 2014). Landscape appropriation refers to the action, material and symbolic, by which human beings extract elements or benefit from ecosystem by converting them into a social element, for example, filtration, retention, and storage of fresh water. Likewise, the appropriation and the constant interactions between humans and natural subsystems drive land use and cover changes (Rindfuss et al. 2008). A land-use change may affect land cover, while changing land cover may similarly affect land use (Zvoleff et al. 2014), leading to land-use transitions, that are any change in land-use systems from one state to another. One example is an annual crop for local consumption being replaced with a large tree plantation, due to new market demands (Lambin and Meyfroidt 2010).

Particularly, the traditional landscape appropriation in the Pacific region of Colombia, since the abolition of slavery in the nineteenth century, has been characterised by the multiple use of different goods and services by Afro-descendants based on a series of ancestral practices and logic. According to Restrepo (1996), it involves complex models of production, which demand detailed knowledge of the environment and successful adaptations allowing people to use different ecosystems to satisfy their basic needs without destroying them. Therefore, these territories are a dynamic mosaic of habitats and land uses, including large extensions of forest, croplands and Afro-descendant settlements that practice a traditional strategy of landscape appropriation at different levels. This can be considered in the framework of socio-ecological production landscapes and seascapes (SEPLS).

In relation to Afro-descendant communities and their territories, the National Policy Constitution of Colombia of 1991, and particularly, Law 70 of 1993, allow the organisation of these communities under the juridical figure of Community Boards (*Consejos Comunitarios* in Spanish), to serve functions that include conservation of their natural environment and their cultural identity. However, the future of many SEPLS in the Pacific region is still uncertain, as they could be in danger of disappearing due to external pressures (Quintero-Angel 2016).

In this complex context of SEPLS in the Colombian Pacific region, transformative change is necessary to maintain human activities within planetary limits and to establish and maintain productive, extractive, and conservation practices which are less disruptive to ecosystems and their inhabitants (Ashley and Plesch 2002).

Therefore, this chapter reviews lessons for transformative change, taking as an example the appropriation of the landscape and the transitions of land use in the community of San Marcos. San Marcos is one of 196 collective territories (eight million hectares in total) of Afro-descendant communities existing in Colombia. These territories are known as Afro-descendant settlements with agriculture, livestock and mining traditions inherited ancestrally from people who were slaves, runaway slaves and free people who looked after runaway settlements (*palenques* in Spanish) usually near rivers (Instituto Colombiano de la Reforma Agraria, resolution 2066 of 2002).

10.1.1 Importance of the Colombian Pacific Region

The Colombian Pacific region is part of the Choco biogeographic region, which is recognised as one of the most biodiverse areas worldwide. It is characterised by lowland rainforest with an exuberant diversity of plants and animals and a high level of endemism compared with other regions worldwide (Plotkin et al. 2000; Losos and Leigh 2004; Rangel et al. 2004). Despite being one of the richest biodiversity areas, the region's human population has one of the highest levels of poverty and social inequity, with incomes inferior to the national rural and urban averages (Barbary et al. 2004). Additionally, pressure on its mineral and forest resources has caused social conflict, which has increased the presence of illegally-armed groups and drug trafficking-related actors (Quintero-Angel 2015b). This situation has put the region's flora and fauna in danger, and has also endangered the communities of Afro-descendants and indigenous people in coastal and riverine zones. These populations enter into conflict with external immigrants and have a high tendency to solve problems in violent ways (Contraloría General de la República 2013).

Lowland rainforest in this region covers around 77% of the land (Escobar 2008). According to Rangel et al. (2004), the Pacific coast region of Colombia has 5474 plant species in 1406 genera and 271 families. In terms of fauna, this region records 134 amphibian species and 166 reptiles in the Pacific lowlands (Velasco et al. 2008), 793 bird species in 447 genera and 73 families (Rangel et al. 2004), and 12 orders of mammal species in 114 genera and 180 species (65% of the total genera and 40% of Colombian species) (Muñoz and Alberico 2004).

The regional soils are characterised by low content of organic matter and a highly acidic pH, which implies a low quality of soil for agriculture (Jaramillo 2002). As observed by West (1957) in the mid-twentieth century, productive soils for agriculture are scarce and limited to narrow bands in the alluvial plains of the Pacific coast region. Escobar (2008) stated that strips along rivers, dams, and meadows of different sizes provide space for human settlements and crops such as corn, coconut, cocoa, and plantain.

10.1.2   The Collective Territory of San Marcos

San Marcos is an ethnic territory of 3689 ha (Instituto Colombiano de la Reforma
Agraria, resolution 2066 of 2002) (Ramírez 2006a) with a current population of
250 inhabitants, who are Afro-descendants of several generations. San Marcos is
located in the rural area of the industrial, port, biodiversity and eco-tourism special
district of Buenaventura, in the low basin of the Anchicaya river. The special district
of Buenaventura is placed on the Pacific slope of the western mountain range, in the
Valle del Cauca Department of Colombia, close to the Farallones de Cali National
Park (Fig. 10.1 and Table 10.1). The region corresponds to a lowland rain forest
biome, with an average temperature of 28 °C and annual average precipitation of
6600 mm (IREHISA 2013). Both San Marcos and the low Anchicaya areas are
highly degraded especially due to deforestation, mechanised surface mining and the
construction of two dams on the Anchicaya river during the twentieth century for
hydropower generation.[1]

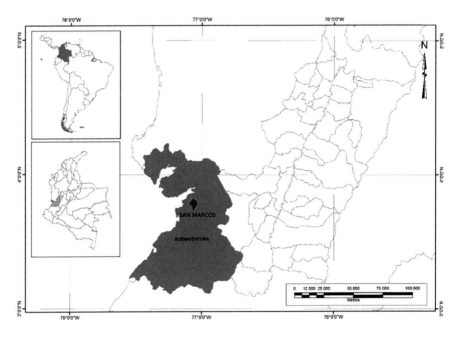

Fig. 10.1 Location of San Marcos

---

[1]The first dam, which began operations in 1955, is located in the low Anchicaya basin with a total
capacity of 64 MW, a median annual power of 360 GWh and a median fall of 72 m. The second dam
is located in the high Anchicaya basin, and started operations in 1974 with an installed capacity of
340 MW, annual median energy of 1590 GWh and a height of 140 m (Larrahondo 1993).

Table 10.1 Basic information of the case study area

Country	Colombia
Province	Valle del Cauca
District	Buenaventura
Municipality	San Marcos
Size of geographical area (hectare)	607,800
Number of indirect beneficiaries	423,927
Dominant ethnicity(ies), if appropriate	Afro-Colombian communities
Size of case study/project area (hectare)	3,689
Number of direct beneficiaries	250
Dominant ethnicity in the project area	Afro-Colombian communities
Geographic coordinates (latitude, longitude)	3° 42′ 23.04″ N; 76° 57′ 32.04″ W

In the ecological zoning of San Marcos, around 93% of the territory is allocated for conservation (especially in middle and high zones), and 7% for productive activities and human settlements (Ramírez 2006b), especially in low zones. Meanwhile, 77% of the territory is collectively owned, while 12.5% is occupancy without legal documents that support the property and 10.5% is titled private property (Ramírez 2006a). According to Pérez (2008), a majority of the soils in Buenaventura (97%) are classified as having low or very-low productivity, which limits their development potential for agricultural activities.

According to a classification of Pacific settlements by Mosquera and Aprile (2006), San Marcos is a fluvial settlement typical of those along the Pacific, where generations of involvement in agricultural production, harvest, and shipping to nearby markets has generated a system whereby some inhabitants deal with transport, exchange and distribution, while the majority work in agroforestry and harvesting activities. The settlement is located in one of the less hilly areas, closest to the San Marcos stream, along the old road between Cali and Buenaventura.[2] Houses are built with wood or brick. The settlement has a school, health centre, several stores and an administrative community centre (Fig. 10.2).

10.2   Methods

An initial literature review on the collective management of the landscape in Afro-descendant territories was carried out, starting from the time of the below-mentioned Law 70 of 1993. Subsequently, the landscape appropriation strategies of the early twenty-first century were identified and categorised using the appropriation flow model proposed by González de Molina and Toledo (2014). According to Toledo

---

[2]Generally, other settlements in the Pacific region of Colombian tend to evolve around the river. However, in San Marcos and low Anchicaya, human settlements were built along the road, generally with some parts parallel to the river.

A.   B.

Fig. 10.2 General aspects of San Marcos: (**a**) Panoramic view. (**b**) Settlement. Photos by M Quintero-Angel

Table 10.2 Description of landscape units of nature appropriation flow model. Adapted from Cordón (2007)

Landscape units	Definition	Types of activities or exchanges
Used Environment (UEN)	Refers to a group of elements (e.g. water, and soils) that are appropriated without causing a break in the ecosystem structure	All known forms of hunting, fishing, collection and sheep husbandry, as well as certain forms of low-impact extraction and animal breeding, such as foraging in original land cover
Transformed Environment (TEN)	Artificial ecosystems resulting from human labour that alter the ecosystem balance	All forms of agriculture, livestock farming, forestry, development of plantations, aquaculture, and mining (high impact)
Conserved Environment (CEN)	Untouched areas that do not offer goods but diverse ecosystem services instead	All activities related to the conservation of ecosystems and ecosystem services
Social Environment (SEN)	Sectors of the society as a whole that fall outside the limits of the P unit, such as carrying out some types of exchange using units of appropriation	Exchanges of an economic type that communities carry out with the larger society, such as trade or exchange of products, paid labour, etc.

(2008), units of appropriation (P) correspond to a group of individuals that enjoy, possess, dominate, or are owners of a fragment of nature which they exploit to subsist by extracting elements from the biosphere (living beings, water, and air) and the geosphere.

For this analysis, all units of rural production/extraction of households in San Marcos correspond to a single unit P, where two types of relationships or exchanges are recognised: (1) interactions of the ecological type with the surrounding environment: the used environment (UEN), the transformed environment (TEN) and the conserved environment (CEN); and (2) interactions of an economic type: social environment (SEN) (Table 10.2). The flow model considers the system of interactions between these four landscape units (UEN, TEN, CEN and SEN) and unit P (González de Molina and Toledo 2014).

Information-gathering techniques included semi-structured interviews and participant observation, which were conducted through several visits between January 2013 and December 2014, during the fieldwork of this study (Quintero-Angel 2015b). The information gathered consisted of the general aspects of life in the community, dynamics of landscape appropriation, and explanatory information on the management of nature. In total, 35 interviews were conducted, 29 of which were carried out with local informants, 11 were recorded and transcribed, and 18 were documented in the field book. Additionally, six interviews were conducted with informants from outside the community, two were recorded and transcribed, and four were documented in the field book. Other researchers (Cordón 2007; Cordón and Toledo 2008; García-Frapolli et al. 2008; Quintero-Angel 2015a, 2016) have used these same ethnographic techniques to study landscape appropriation.

10.3 Results and Discussion

10.3.1 Management of Afro-Descendant Collective Territories from Law 70 of 1993

In 1991, the Colombian government organised a National Constituent Assembly to reform the Political Constitution of 1886. The new Political Constitution highlights the inclusion of fundamental and third-generation rights such as economic, social, and collective rights, that cover respect and guarantees for minority ethnic groups. These ethnic groups (Afro-descendants, indigenous and Romani people) historically did not have any social, cultural, environmental or territorial guarantees (Sánchez et al. 1993).

Particularly, the transitory article 55 of the Constitution of 1991 established the right of the Afro-descendant communities in the Pacific region to collective ownership of land (Leal 2008).[3] Then on 27 August 1993, Law 70 of 1993 was issued. The purpose of this law was to recognise the collective ownership right of Afro-descendant communities[4] who had occupied[5] empty lands in riverine rural areas in the Pacific basin, following traditional practices of production (Article 1, Law 70 of 1993). Traditional practices refer to agriculture, manual mining, forest extraction, livestock husbandry, hunting, fishing and other harvesting of natural products, which

---

[3]Paragraph 1 of Article 1 makes it clear that the law targets not only the populations of the Pacific Basin, but all communities with similar conditions in other areas as well.

[4]An Afro-descendant community refers to a group of families of Afro-Colombian descendant who has its own culture, shares a history and has its own traditions and customs within a rural-urban setting, and who reveals and conserves a conscience of identity that differentiates from other ethnic groups (Article 2, Law 70, 1993).

[5]The law refers to collective occupation, meaning the historic and ancestral occupation of Afro-descendant communities on land for their collective use, the land constituting their habitat, on which they developed their traditional practices of production (Article 2, Law 70, 1993).

have been used to guarantee the conservation of their lives and autonomous sustainable development (Article 2, Law 70, 1993). Additionally, this law looks to establish mechanisms to protect the cultural identity of the Afro-descendant communities of Colombia as an ethnic group and to support their economic and social development to guarantee that these communities obtain tangible conditions of equality compared with the rest of the Colombian society (Article 1, Law 70 of 1993).

Before Law 70 of 1993, the Colombian state, as owner of the lands in rural areas in the Pacific region, dealt with them without considering the ethnic groups inhabiting the territory. Thus, the government facilitated and promoted the arrival of extractive entities and granted licences to companies for timber extraction and gold and platinum mining, among other activities, which degraded many SEPLS in the Colombian Pacific.

Law 70 of 1993 established that collective occupation of a territory is eligible for a collective title, meaning that historic occupations could then be recognised as property rights, calling the territories "Afro-descendant community lands". However, to get access to the collective entitlement of the territory, it was necessary for the communities to organise administrative units called "community boards". Some of the functions of the community boards include the delimitation and assignment of areas in the entitled land, conservation and protection of collective property rights and preservation of cultural identity, as well as use and conservation of natural resources, among others (Article 5, Law 70 of 1993). The requirement to form this community board to obtain access to the collective title and to act as a new organisation for the political and territorial management was a key feature of Law 70. Key decisions need to be discussed in a general assembly constituted by all inhabitants in the collective territory, who elect a community board for 3 years (Vélez 2011).

The application of this law for Afro-descendant communities has been complex, as these communities are diverse and heterogeneous. This law considered all Afro-descendant communities as the Pacific Region communities model, which denied the diverse levels of appropriation that Afro-descendant communities had given to inter-Andean valleys in the Colombian south and Caribbean (Duarte and Rodríguez 2014). There was also conflict in the formation or organisation of Afro-descendant groups in the rush to take advantage of the instrumental use of the Law, which led to some community boards being organised based on weak political significance and ethnic affirmation (Agudelo 2005; Romaña et al. 2010). Another situation that damaged the application of the Law was the location of many community boards with the presence of illegally-armed groups leading to human rights violations, e.g. forced displacement, forced disappearances, extrajudicial disappearances, and murder of social leaders.

In general, it has been recognised that the collective entitlement promoted by Law 70 did not have enough public funds to finance monitoring of local authorities, implementation of local management plans (Vélez et al. 2020), and the application of the law completely, as "there is no development of decrees or articles, for example, that allows for their economic autonomy or a major incidence in aspects

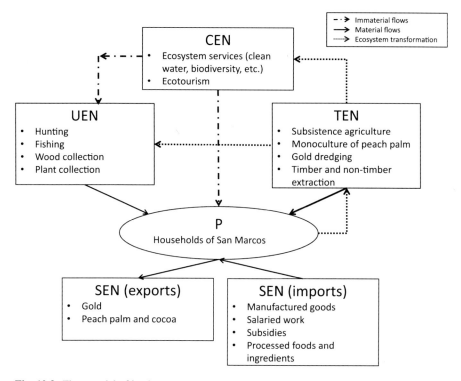

Fig. 10.3 Flow model of landscape appropriation in San Marcos

like health or education" (Duarte and Rodríguez 2014, p. 4). Additionally, others have noted that the collective entitlement promoted by Law 70 cannot completely assuage the pressures of the illegal economy, including coca crops and gold mining. In general, these extractive activities are associated with illegally-armed actors that infiltrate the communities through violence and interrupt the capacity of local organisations to comply with rules and norms and modify traditional practices for landscape appropriation (Vélez et al. 2020).

10.3.2   Landscape Appropriation in the San Marcos SEPLS

The diversity of productive and extractive activities performed in the framework of the landscape appropriation in San Marcos at the beginning of the twenty-first century is represented in a model of goods and services flow (Fig. 10.3). The flow model indicates that the San Marcos community has developed a multiple-use

A   B

Fig. 10.4 Traditional practices in San Marcos: (**a**) The traditional way of transporting the harvest of peach palm (*Bactris gasipaes*), (**b**) Traditional elaboration of a wooden pan for artisanal mining. Photos: M Quintero-Angel, 2015

strategy that combines time and spaces (in UEN, TEN, and CEN): multiple crops under the slash and decompose system[6] grown dispersed without a pattern between wild species in the river margins and alluvial plains, with plant species such as papachina (*Colocasia esculenta*), banana (*Musa paradisiaca*), borojo (*Borojoa patinoi*), and peach palm (*Bactris gasipaes*); fishing; hunting with traps oriented towards self-supply of the household; firewood collection; wood collection for building material; other forest fruit collection; domestic crafting; and artisanal extraction of gold, among others (Fig. 10.4). It is a multiple-use strategy that looks for diversification of activities in the landscape to obtain all materials, energy, water, and services required, relying more heavily on the UEN (González de Molina and Toledo 2014). This strategy puts pressure on different natural elements in time and space, denoting an ecological local knowledge and high adaptation to the environment.

Multiple-use strategies have their origin in the traditional or ancestral practices of the Afro-descendant cosmovision, which controls or regulates human activity and/or promotes solidarity. According to Escobar (2008), the natural world had an intimate presence in the cultural imagination of Afro-Colombian groups, which is elaborated in their narratives and shows that African, indigenous and catholic traditions have transmitted an ecological ethic of reciprocity and conservation, serving to advise people not to abuse nature. For example, through some myths and mythological characters, people are advised to stay away from certain areas of the forest or just to hunt or fish what is needed to feed their households. The multiple-use strategy is also supported by solidarity, reflected in the practice of sharing agricultural, hunting and

---

[6]The slash and decompose method (*tumba y pudre* in Spanish) consists of cutting weeds around the plants of interest. The weeds are then left to rot on the ground, contributing organic matter to very mineral-poor soils.

fishing products in exchanges, and the practice of "hand change"[7], or collective work, on farms.

The traditional multiple-use system in San Marcos shares important features with different agroforestry strategies identified by Alcorn (1990) in Latin America, in that it: (1) incorporates native species; (2) uses natural variations of the environment; (3) makes use of natural succession to generate resources, protect the soil and control pests; (4) incorporates a large number of species; (5) develops flexible and individualised strategies; (6) reduces risks with diversification; and (7) seeks to ensure that independence survives.

In the Colombian Pacific region from the second half of the twentieth century, pressures from SEN altered the multiple-use strategy of appropriation in San Marcos (Escobar and Pedrosa 1996; Escobar 2008). On the one hand, some households transformed the multiple strategies to incorporate other activities that provide a diverse source of economic income through paid work, sale of products, rent, monetary remittance from relatives, subsidies, and return on investment outside San Marcos (e.g. renting a house in the city), among others. On the other hand, other households of San Marcos were pressured towards specialisation, such as selling one agricultural product, mechanised mining or paid work in nearby cities.[8] Agricultural production, since the mid-1980s, presented an intensification of peach palm fruit farming, stimulated by the government, NGOs and private companies, who sent agronomists and other professionals to advise the community on the agricultural practices under a monoculture system and technological packages of the Green Revolution (Fig. 10.5a). The professionals taught the community that traditional practices are rudimentary and must be changed into monoculture. This production generated new paid jobs, daily work cleaning plots, fumigating or harvesting in exchange for money.[9] Mechanised mining is an illegal activity of surface gold extraction, which is performed without environmental licence or restoration plan. The surface mining is performed with excavators controlled by people outside of the community who rent the land occasionally, incurring violence and/or unequal distribution of benefits, whereby the owner of the excavator receives 90% of the income, and the community only 10% (6% for the landowner and 4% for the community board), as well as 100% of the significant negative environmental impacts (Quintero-Angel 2015a) (Fig. 10.5b).

In general, the specialised strategy brought in external people, illegal exploitation of soil, loss of land cover, changes in land use, reduction in species diversity, loss of agricultural soils, landscape transformation, water contamination, traditional job

---

[7]Collective and free work on farms or in forest for some benefit, which rotates, generating a relationship of reciprocal benefit. In San Marcos, two or more people assist each other for 1 day on a farm, then one of them the next day on the farm of another, and so on. During fieldwork in 2013 and 2014, it was possible to establish that this practice is still conserved by some male elders.

[8]In the twentieth century, up to the 1990s, the specialised strategy in San Marcos was focused on timber harvesting for the commercial market.

[9]These agricultural practices are not accounted for in formal contracts, nor is there coverage for work-related risks.

A

B

Fig. 10.5 (**a**) Peach palm monoculture farm. (**b**) Significant negative environmental impacts associated with the removal of vegetation cover and alteration of soil layers due to turn over for gold extraction by mechanised mining in San Marcos. Photos: M Quintero-Angel, 2014

substitution, violence for the control of the land, and forced displacement of families. Additionally, it led to cultural fragmentation which corresponded to changes in relationships with nature, incorporation of new practices, as well as transformations in the used technology, knowledge, cosmovision and institutions or norms that had immersed the population in a severe crisis at the beginning of the twenty-first century (Quintero-Angel 2015b).

Between 2013 and 2014, mechanised mining and peach palm crops generated a severe crisis for households that still continue today. Peach palm crops were decimated by pests (*Dynamis borassi* and *Rhynchophorus palmarum* beetles). The government then took control of illegal mining by having the army blow up all excavators present in the lower Anchicaya region. A member of the community board described the situation like this: "...now there is nothing to do, the few incomes here are from the mine, there is no peach palm, what I mean is that in this moment we are in a critical situation" (interview with a member of the community board, 2014).

10.3.3   Lessons for Transformative Change

According to Toledo (2003), rural communities should maintain a hybrid character, combining the achievements of tradition with the positive aspects of modern society. However, replacement of the multiple-use strategies by a specialised ones in San Marcos has meant this has not been successfully carried out.

In general, the collective entitlement of Law 70 of 1993 reduced deforestation rates significantly, but its effect varied substantially depending on the subregion, organisation of the community to define rules for community use of the natural resources, expulsion of private companies dedicated to wood extraction and palm oil plantations, as well as the resistance against illegal groups and their pressures towards mechanised mining and drug trafficking (Vélez et al. 2020).

However, the responsibility for the actions for transformative change cannot be placed only on the communities and community boards.

> As one leader from Nariño mentioned, "the Councils have struggled here, but the intimidation has not allowed the impact to be successful..... people have died here for standing up against certain dynamics: coca, oil palm... the councils have resisted..., but it is not enough" (Vélez et al. 2020, p. 12).

For transformative change, the Colombian government must play the main role in supporting the communities and recognising the socio-ecological context, so that the local government, NGOs, and other actors can improve local productive and extractive practices while maintaining local values. Also, the government must confront illegal activities and not allow intimidation of community councils (i.e. Afro-descendant collective territories under Law 70) and communities. Moreover, it should develop "new instruments to control illegal markets since the definition of property rights in the context of a weak state without the monopoly of violence is insufficient to protect the forest" (Vélez et al. 2020, p. 12). Also, it is essential that the government recognises and promotes the conservation of multiple-use strategies and the development of business activities compatible with traditional practices and ecosystem services offered by the territory. For example, a scheme for productive vocations utilising ecosystem services could be fomented by marketing products derived from the biodiversity, oriented towards markets with high aggregated value. Current policies based on monoculture implemented in recent years need to be reviewed.

To recognise, revalue and recover the Afro-descendant multiple-use strategy oriented towards conservation of SEPLS, it is necessary to develop ethno-education strategies. Ethno-education is "offered to groups or communities comprising the nationality who possess a culture, a language, traditions and their own and autochthonous status (. . .[, and]) should be linked to the environment, productive process, and social and cultural process, with respect towards their beliefs and traditions" (Law 115 of 1994, National Ministry of Education). It constitutes an instrument for the transmission of knowledge and ancestral wisdom, which is at risk of disappearing, to the younger generation, who will have the future of the communities in their hands.

In this sense, it is necessary to establish a new governance scheme with a fair balance between the State, civil society, and the economy of local Afro-descendant communities, based on harmony with nature. This new governance scheme must be integrative, adaptive, informed and inclusive, with a real and effective application of Law 70 and taking into account leverage points that guarantee the application of traditional multiple-use practices of these communities, as well as the incorporation of new technologies that respect the environment and cultural traditions. The different actors and their roles in actualising transformative change are listed in Table 10.3.

Finally, the appropriation flow model used to study landscape appropriation in San Marcos (Fig. 10.3), may be useful to assess or monitor transformative change in the Colombian Pacific region. In this regard, some indicators of landscape appropriation can be set as those proposed by Cordón (2007), e.g. the number of productive/extractive activities and the number of species used in each landscape unit. Likewise, other indicators could be used, such as the number of productive/extractive activities aimed at household self-sufficiency and the market, or total area and/or ratio of land surface dedicated by members of the community council to the multiple-use strategy, among others. Using such indicators, the actors involved in the governance of Afro-descendant territories may monitor and evaluate the impact of their interventions over time.

10.4   Conclusions

The multiple-use strategy of landscape use is flexible over time and space, giving more resilience to each household,[10] and in turn giving the community greater capacity to adapt to global changes using diversification as a strategy of subsistence, focused on different natural elements across time. However, in San Marcos, and generally in the Colombian Pacific region, there is a conflict between conservation and development, which is currently pressuring communities toward specialisation of household subsistence strategies in the production and sale of one agricultural product and/or paid labour in nearby cities. This strategy focused on specialisation signifies a major risk to communities that face changes in demand for products, price and environmental conditions, which are variable in nature and altered by climate change.

In San Marcos, the transition from a multiple-use strategy towards more specialised use signified changes in beliefs system and local knowledge, transforming the ecological ethic of reciprocity and conservation based on the

---

[10]In the context of this research, resilience essentially includes three aspects: (1) quantity of change that the system can experience while still maintaining the same control over its function and structure; (2) degree in which the system is able to auto-organise; and (3) skills and increased capacity for learning and adaptation. Adapted from Cabell and Oelofse (2012).

Table 10.3 Description of actions required for transformative change for different actors present in San Marcos

Actor	Actions required for transformative change
Colombian national government	• Monitoring of compliance with legislation and protection of the rights of community councils • Allocation of resources for application of Law 70 of 1993 • Provide guarantees to stop illegal activities in Afro-descendant territories • Promotion of ethno-education • Organisational strengthening for citizen participation in community councils • Promote and fund research on the multiple uses of landscape strategy • Promote productive strategies that maintain the strategy of multiple uses of landscape. e.g. ethno-tourism
Afro-descendant communities (community councils)	• Commitment to transfer traditional knowledge on the use of the landscape from generation to generation • Support and promotion of ethno-education strategies • Strengthening and promotion of the multiple-use strategy of the landscape • Diversified productive/extractive activities based on the multiple-use strategy of the landscape, e.g. promoting ethno-tourism • Coordination and planning of landscape use • Organisational strengthening for better dialogue with the State and to improve the governance of its territories • Use legal mechanisms of citizen participation provided in the Colombian Constitution to enforce the State to comply with the implementation of Law 70 of 1993
Educational and research institutions	• Study and support the application of the multiple-use strategy • Design of indicators for monitoring the multiple-use strategy and the traditional systems • Guide and support the design of strategies of community councils for ethno-education, heritage conservation, and planning, among others • Management of the establishment of strategic alliances between community councils, companies, and the State
Companies	• Technical assistance adapted to the socio-cultural context of the communities • Technical assistance respectful of the local culture • Financing of actions for the conservation of the cultural heritage associated with the use of the landscape
International and regional organisations	• Financing and support for local development programs based on improvement of the landscape and multiple-use strategy • Financing of actions for the recovery of traditional knowledge • Cooperate with the Colombian government for the conservation of the multiple-use strategy and to stop illegal activities in Afro-descendant territories

Afro-descendants' traditional rational use of nature into a more utilitarian vision. Therefore, in collective territories like San Marcos, reinforcement of mechanisms that bring together the community and individual interests need to be promoted. Additionally, public policies to generate new employment and income for locals, thereby causing less negative environmental impacts, need to be introduced.

For the future of San Marcos, it is very important to recognise, revalue and recover Afro-descendant strategies of multiple use, which contribute to biological and cultural conservation in the Colombian Pacific region. Science, technology, and innovation should be oriented towards deepening and improving the multiple-use strategies of the communities and offering alternatives and information to maintain practices of fishing, hunting, collection and gold mining within a framework of the productive limits of these ecosystems. These activities should involve community labour and the training of leaders, combined with development based on ancestral knowledge that reinforces the conservation of agrobiodiversity, strengthening the value to the people of continuing to live in the region. Also, an ethno-education strategy of knowledge divulgation wherein young people learn and appreciate ancestral practices is required.

The challenge in San Marcos, and other SEPLS along the Pacific coast of Colombia, will be to find sustainable ways of landscape appropriation that not only allow for the conservation of ecosystems, but also for the well-being of the human population. Accumulated local knowledge and the creativity of Afro-descendants are not only fundamental, but also represent the best hope to improve the community's future and become key to transformative change that will enable the conservation of the SEPLS.

References

Agudelo, C. (2005). *Retos del Multiculturalismo en Colombia, Política y Poblaciones Negras.* Medellín: La Carreta Social.

Alcorn, J. B. (1990). Indigenous agroforestry systems in the Latin American tropics. In M. A. Altieri & S. B. Hecht (Eds.), *Agroecology and small farm development* (pp. 203–218). Boca Ratón: CRC Press.

Ashley, K., & Plesch, V. (2002). The Cultural Processes of "Appropriation". *Journal of Medieval and Early Modern Studies, 32*(1), 1–15.

Barbary, O., Ramirez, H. F., Urrea, F., & Lopez, C. A. V. (2004). Perfiles contemporáneos de la población afrocolombiana. In O. Barbary & F. Urrea (Eds.), *Gente negra en Colombia: Dinámicas sociopolíticas en Cali y el Pacífico.* Colombia: CIDSE, IRD, COLCIENCIAS.

Cabell, J. F., & Oelofse, M. (2012). An indicator framework for assessing agroecosystem resilience. *Ecology and Society, 17*, 18.

Contraloría General de la República. (2013). *La explotación ilícita de recursos minerales en Colombia: casos Valle Del Cauca (río Dagua) – Chocó (río San Juan) efectos sociales y ambientales*, Sistema General de Regalías, viewed 28 May 2020, https://www.contraloria.gov.co/documents/20181/198738/Separata-Mineria-Ilegal.pdf/4d3d5cbe-4bda-430a-831e-ef2f6bbf5d0d?version=1.0

Cordón, M. R. (2007). *Estrategias Indígenas, Conservación y Desarrollo Rural Sostenible en la Reserva de la Biosfera de Bosawas, Nicaragua, Programa de Agroecología, Sociología y Desarrollo Rural Sostenible.* Córdoba: Universidad de Córdoba.

Cordón, M. R., & Toledo, V. M. (2008). 'La importancia conservacionista de las comunidades indígenas de la Reserva de Bosawás, Nicaragua: un modelo de flujos. *Revista Iberoamericana de Economía Ecológica, 7*, 43–60.

Duarte, C., & Rodríguez, T. (2014). Los desafíos de la sostenibilidad rural colombiana en un escenario de pos conflicto, equidad, diferencia y reciprocidad participativa. In *4to encuentro internacional sobre interculturalidad: Territorialidades, desarrollo rural y paz*. Cali: Pontificia Universidad Javeriana.

Escobar, A. (2008). *Territories of difference: place, movements, life, Redes*. Durham: Duke University Press.

Escobar, A., & Pedrosa, A. (1996). Conclusión: Globalización, posdesarrollo y pluriculturalismo. In A. Escobar & A. Pedrosa (Eds.), *Pacífico ¿Desarrollo o diversidad?: Estado, capital y movimientos sociales en el Pacífico colombiano*. Bogotá: CEREC-ECOFONDO.

García-Frapolli, E., Toledo, V. M., & Martinez-Alier, J. (2008). Adaptations of a Yucatec Maya multiple-use ecological management strategy to ecotourism. *Ecology and Society, 13*(2), 31.

González de Molina, M., & Toledo, V. M. (2014). *The Social Metabolism: A socio-ecological theory of historical change*. New York: Springer International Publishing.

IREHISA. (2013). *Análisis exploratorio y confirmatorio de datos de la cuenca del río Dagua, informe Proyecto, Identifying signs of climate variability and change in an Andean Basin and Mediterranean, to define adaptation strategies for the management of water resources*. Cali: Universidad del Valle.

Jaramillo, D. F. (2002). *Introducción a la ciencia del suelo*. Medellín: Universidad Nacional de Colombia.

Lambin, E. F., & Meyfroidt, P. (2010). Land use transitions: Socio-ecological feedback versus socio-economic change. *Land Use Policy, 27*(2), 108–118.

Larrahondo, M. (1993). Aprovechamiento acuícola de embalses en Colombia. In J. R. Juárez & E. Varsi (Eds.), *Avances en el manejo y aprovechamiento acuícola de embalses en América Latina y el Caribe*. México: Organización de las Naciones Unidas para la Agricultura y la Alimentación – FAO, viewed 28 May 2020, http://www.fao.org/docrep/field/003/ab488s/AB488S00.htm#TOC.

Leal, C. (2008). Disputas por tagua y minas: recursos naturales y propiedad territorial en el Pacífico colombiano, 1870-1930. *Revista Colombiana de Antropología, 44*, 409–438.

Losos, E., & Leigh, F. G. (2004). *Tropical forest diversity and dynamism: Findings from a large-scale plot network*. Chicago: University of Chicago Press.

Mosquera, G., & Aprile, J. (2006). *Aldeas de la costa de Buenaventura, Hábitats y sociedades del pacífico*. Cali: Programa Editorial Universidad del Valle.

Muñoz, Y., & Alberico, M. (2004). Mamíferos en el Chocó biogeográfico. In J. O. Rangel (Ed.), *Colombia Diversidad Biótica IV. El chocó Biogeográfico/Costa Pacífica* (pp. 559–598). Bogotá: Universidad Nacional de Colombia, Instituto de Ciencias Naturales, Conservación Internacional.

Pérez, G. J. (2008). Historia, geografía y puerto como determinantes de la situación social de Buenaventura. In J. V. de la Hoz (Ed.), *Economías del Pacífico colombiano* (pp. 51–81). Cartagena: Banco de la República.

Plotkin, J. B., Potts, M. D., Yu, D. W., Bunyavejchewin, S., Condit, R., Foster, R., et al. (2000). Predicting species diversity in tropical forests. *Proceedings of the National Academy of Sciences, 97*, 10850–10854.

Quintero-Angel, M. (2015a). Aproximación a la racionalidad ambiental del extractivismo en una comunidad afrodescendiente del Pacífico colombiano. *Revista Luna Azul, 40*, 154–169.

Quintero-Angel, M. (2015b). *Dinámica socio-ecológica del uso y transformación de la naturaleza en San Marcos (Buenaventura-Colombia)*. Valle del Cauca: Facultad de Ciencias, Universidad del Valle.

Quintero-Angel, M. (2016). Apropiación de la naturaleza por una comunidad afrodescendiente del pacífico colombiano: un modelo de flujos. *Revista Iberoamericana de Economía Ecológica, 25*, 1–15.

Ramírez, M. (2006a). *Mapa de usos del territorio*. Buenaventura: Corporación Autónoma Regional del Valle del Cauca.

Ramírez, M. (2006b). *Mapa político administrativo*. Buenaventura: Corporación Autónoma Regional del Valle del Cauca.

Rangel, J. O., Aguilar, M., Sanchez, H., & Lowy, P. (2004). Región Costa Pacífica. In J. O. Rangel (Ed.), *Colombia Diversidad Biótica I* (pp. 121–139). Bogotá: Instituto de Ciencias Naturales-Universidad Nacional de Colombia-Inderena.

Restrepo, E. (1996). Cultura y Biodiversidad. In A. Escobar & A. Pedrosa (Eds.), *Pacífico ¿Desarrollo o diversidad?: Estado, capital y movimientos sociales en el Pacífico colombiano* (pp. 220–244). Bogotá: CEREC-ECOFONDO, viewed 28 May 2020, http://www.ram-wan.net/restrepo/documentos/culturaybiodiversidad.pdf.

Rindfuss, R. R., Entwisle, B., Walsh, S. J., An, L., Badenoch, N., Brown, D. G., et al. (2008). Land use change: complexity and comparisons. *Journal of Land Use Science, 3*(1), 1–10.

Romaña, N., Bonilla, C., Zapata, F., & Gonzalez, E. (2010). *Titulación Colectiva para comunidades negras en Colombia*. Bogotá: Espacio Creativo Impresores.

Sánchez, E., Roldán, R., & Sánchez, M. (1993). *Derechos e identidad: los pueblos indígenas y negros en la Constitución política de Colombia de 1991*. Bogotá: Disloque Editores.

Toledo, V. M. (2003). Hacia un modelo de conservacion bio-regional en las regiones tropicales de mexico: biodiversidad, sustentabilidad y pueblos indigenas. In *Hacia una Evaluación de las Áreas Naturales Protegidas del Trópico*. Xalapa: Universidad Veracruzana, viewed 28 May 2020, http://www.era-mx.org/biblio/politica/Toledo2003.pdf.

Velasco, J. A., Quintero-Ángel, A., & Garcés, M. F. (2008). Diversidad de especies de anfibios y reptiles en las tierras bajas del Pacífico del Valle del Cauca. *Cespedesia, 31*(86–87), 81–94.

Vélez, M. A. (2011). Collective titling and the process of institution building: the new common property regime in the Colombian Pacific. *Human Ecology, 39*, 117–129.

Vélez, M. A., Robalino, J., Cardenas, J. C., Pas, A., & Pacay, E. (2020). Is collective titling enough to protect forests? Evidence from Afro-descendant communities in the Colombian Pacific region. *World Development, 128*, 104837.

West, R. (1957). *Las tierras bajas del Pacífico colombiano*. Bogotá: Icanh.

Zvoleff, A., Wandersee, S., An, L., & López-Carr, D. (2014). *Land use and cover change*. Oxford: Bibliographies.

Laws and Regulations

Instituto Colombiano de la Reforma Agraria, resolution 2066 of 2002. Diario Oficial de Colombia (45.042), págs. 6-8.

Law 115 of 1994 - Ley 115 de 1994. Diario Oficial de la República de Colombia, Bogotá, Colombia, febrero 8 de 1994.

Law 70 of 1993- Ley 70 de 1993. Diario Oficial de la República de Colombia, Bogotá, Colombia, 27 de agosto de 1993.

Political Constitution of Colombia - Constitución política de Colombia [Const.] 1991, artículo transitorio 55. [Título I, Cap. 2].

Toledo, V. M. (2008). Metabolismos rurales: hacia una teoría económica-ecológica de la apropiación de la naturaleza. *Revista Iberoamericana de Economía Ecológica, 7*, 1–26.

Chapter 11
Climate Change Resiliency Through Mangrove Conservation: The Case of Alitas Farmers of Infanta, Philippines

Dixon T. Gevaña, Josephine E. Garcia, Clarissa D. Ruzol, Felisa L. Malabayabas, Liezl B. Grefalda, Elizabeth O'Brien, Elsa P. Santos, and Leni D. Camacho

Abstract Transformation, transition, and paradigm shift are increasingly applied concepts in literature on climate resiliency to describe changes in society and the environment. Here, we considered mangroves to be dynamic socio-ecological systems, subject to increasing anthropogenic pressures that present complex challenges for the design of effective coastal governance. Analysing these systems through a participatory approach, we consulted a community who lives in close relationship with mangroves, the Alitas farmers of Infanta, Quezon Province, Philippines. This community has improved decision-making processes for the development of adaptation strategies to climate change. In turn, a sustainably managed and conserved mangrove forest promotes human well-being and resilience, particularly for those households whose livelihoods are dependent on the resources that mangroves provide. This paper examined the importance of mangrove land management that addresses climate change hazards. We synthesised various perspectives on the importance of mangrove conservation for enhancing climate resiliency by: (1) describing the climate-related hazards that affect local communities and mangroves; (2) describing socio-institutional structures influencing effective mangrove conservation and local resilience; and (3) identifying climate change adaptation strategies that promote local development and mangrove conservation. This paper establishes a collaborative management framework for future risk-resilience management that operates alongside coastal communities within the Philippines and across the global stage.

D. T. Gevaña (✉) · J. E. Garcia · C. D. Ruzol · F. L. Malabayabas · L. B. Grefalda · E. P. Santos · L. D. Camacho
Department of Social Forestry and Forest Governance, College of Forestry and Natural Resources, University of the Philippines Los Baños, Los Baños, Laguna, Philippines
e-mail: dtgevana@up.edu.ph

E. O'Brien
Department of Ecology and Evolutionary Biology, University of Michigan, Ann Arbor, MI, USA

M. Nishi et al. (eds.), *Fostering Transformative Change for Sustainability in the Context of Socio-Ecological Production Landscapes and Seascapes (SEPLS)*,
https://doi.org/10.1007/978-981-33-6761-6_11

Keywords Adaptation · Climate change · Coastal · Collaborative management · Mangroves · Rehabilitation · Resilience · Stakeholders · Transformative change

11.1 Introduction

Promoting climate risk-resilient coastal communities requires effective anticipation of climate change impacts on the integrity of socio-ecological systems. There is now an increasing need to develop participatory approaches to coastal resilience assessment to improve local understanding of risk and resilience, provide capacity-building benefits, and create platforms for knowledge and experience sharing (Frankenberger et al. 2013; Pfefferbaum et al. 2014; Sharifi and Yamagata 2016). Such assessments are most effectively done by gathering knowledge from a wide range of stakeholders. Designing guidelines on how coastal communities can manage and sustainably use mangrove resources constitutes an important step towards mitigating the potential impacts of climate change.

A participatory approach to coastal resilience assessment is vital in decision-making on trade-offs, as well as fostering local ownership and legitimacy. Such steps are aimed at enhancing the impacts of development services on rural areas through more resilient ecosystems and adaptable communities. By prioritising ecosystem and community resiliency in rural coastal communities, vulnerabilities can be reduced, both over the long and short terms (Renaud et al. 2013). A participatory approach also supports the National Convergence Initiative (Republic of the Philippines, Department of Agrarian Reform), which integrates people, their economy, and the environment, particularly through landscape and seascape approaches.

The concept of resilience has received widespread attention in the field of disaster risk management (DRM) since the release of the "Hyogo Framework for Action 2005–2015: Building the resilience of nations and communities to disasters" by the United Nations International Strategy for Disaster Reduction (UNISDR 2007). This concept is most commonly expressed as the capacity of a society to 'bounce back', or to cope, withstand, resist and recover rapidly from the impacts of hazardous events (IPCC 2012; Ostadtaghizadeh et al. 2015; Turnbull et al. 2013). In the field of climate change, resilience is defined as the ability of a system, community or society exposed to hazards to resist, absorb, accommodate to and recover from the effects of a hazard in a timely and efficient manner (UNISDR 2007).

Resilient coastal communities take deliberate action to reduce risks from coastal hazards with the goal of avoiding disaster and accelerating recovery in the event of a disaster. They adapt to changes based on experience and the applying of lessons learned. Enhancing a coastal community's resilience requires integrating and maintaining an optimal balance of three community-based frameworks, typically viewed as independent and separate domains: community development, coastal management, and disaster management (Fig. 11.1). Community development provides the enabling governance, as well as the socioeconomic and cultural conditions for resilience (Center for Community Enterprise 2000). Coastal management

Fig. 11.1 Resilience as an integrating framework for community development, coastal management, and disaster management domains (Source: US Indian Ocean Tsunami Warning System Program 2007)

provides the framework for managing human use of coastal resources, including mangroves and the coastal zone, in order to maintain environmental and ecosystem resilience (Chua 1998; White et al. 2005). Lastly, disaster management focuses on preparedness, response, recovery and mitigation to reduce human and structural losses from disaster events (Abarquez and Murshed 2004).

Mangrove forests are vital tropical coastal ecosystems that promote resilience. Globally, they offer numerous services that improve the lives of the communities they border (Garcia et al. 2014; Gevaña et al. 2018; Renaud et al. 2013). Mangroves are assemblages of tropical salt-tolerant or halophytic plants that grow between land and sea. Across the Philippines, they have aided in coastal protection during storm surges (Garcia et al. 2014), contributed to local economies (Menéndez et al. 2019), supported marine biodiversity (Honda et al. 2013), and mitigated climate change through carbon sequestration (Camacho et al. 2011; Gevaña and Im 2016; Gevaña et al. 2019). Combined, these services represent the resiliency of the mangrove ecosystem. This ecosystem resiliency can in turn support community resiliency, and vice versa.

Promoting local resilience necessitates an in-depth look at the socio-ecological dimensions of both local communities and mangroves. These dimensions include: local knowledge and appreciation of the socio-ecological production systems of mangroves; commitment to forest conservation; concrete policies, plans and programmes on disaster risk management and climate change adaptation; institutional capacity to enforce and implement these; and mechanisms for equitable sharing of responsibilities and benefits among stakeholders. Pal et al. (2019) further underscores the importance of identifying and nurturing local initiatives that create transformative socio-ecological changes. These transformative changes should be examined in terms of:

- *Enabling environment:* political will; evidence and research/information; and awareness and capacities.
- *Transformational domains:* policies and governance, innovation; and social and behavioural change.
- *Characteristics of transformation:* inclusive development; systemic solutions; catalytic action; ability for scale-up; and sustainability (Mustelin and Handmer 2013).

It is important to emphasise the implementation of participatory approaches in the development and implementation of coastal resilience assessment because of their multiple benefits. According to Sharifi (2016), emphasis should be given to promotion of participatory approaches in all stages of assessment. A number of studies have noted a mix of participatory approaches, such as focus group discussions (FGDs), social mapping, Venn diagrams, resource mapping, institutional relations through mapping, or climate hazards mapping to incorporate local communities' lessons learned and knowledge in results (Akter 2015; Ali et al. 2018; Mallick 2013; Moles et al. 2014; Sameen 2018).

The objective of this paper is to underscore the vital role of multi-stakeholder participation in comprehensive planning and implementation towards strengthening coastal community resiliency. A case study was carried out in Barangay Alitas, Infanta, Quezon, a rural coastal community that is remotely located away from centres of commerce and vulnerable to various climate variabilities. The primary objective of this study was to elicit perspectives on the importance of mangrove conservation for enhancing resilience to climate change hazards. It involved: (1) description of major climatic hazards and events that have affected the local community; (2) analysis of the importance, influence and interests of various mangrove stakeholders; (3) description of major socio-ecological production systems; (4) examination of the local government's institutional capacity to uplift the community's climate resilience; and (5) discussion on the importance of collaborative mangrove management in promoting socio-ecological sustainability.

11.2 Methodology

11.2.1 Description of the Study Area

Alitas is an estuarine *barangay*[1] in the municipality of Infanta in the province of Quezon (Fig. 11.2). It is situated at approximately 14° 42′ 19.44″ N, 121° 40′ 25.32″ E on the island of Luzon, with a total area of 676 hectares (Fig. 11.2 and Table 11.1).

---

[1] A *barangay* is a local community that is recognised as the smallest political unit in the Philippines.

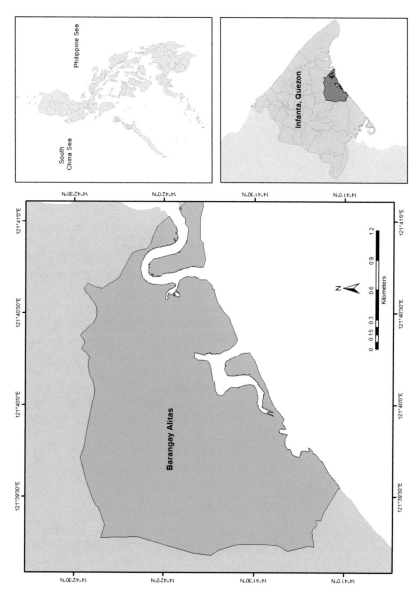

Fig. 11.2 Location map of Barangay Alitas, Infanta, Quezon (Source: PSA and PhilGIS)

Table 11.1 Basic information of the study area

Country	Philippines
Province	Quezon
District	1st district
Municipality	Infanta
Size of geographical area (hectare)	34,276
Number of indirect beneficiaries	69,079
Dominant ethnicity(ies), if appropriate	Tagalog
Size of case study/project area (hectare)	Alitas community (676)
Number of direct beneficiaries	1124
Dominant ethnicity in the project area	Tagalog
Geographic coordinates (latitude, longitude)	14° 42′ 19.44″ N; 121° 40′ 25.32″ E

Its population in 2017 was 1324, which made them one of the less populated *barangays* of Infanta (PSA 2018). The annual population growth rate stands at 0.85%. At this rate, the population is expected to double in 30 years, putting pressure on forestlands and coastal resources via demand for more areas for cultivation, resource extraction, or settlement. Agriculture and fisheries are the common forms of livelihood. Agricultural lands are private areas, while fishery grounds are the mangroves which are of limited access to surrounding communities. The people of Alitas have to rely on multiple sources of income to counter the effects of storms (at least 20 storms pass through the town of Infanta every year), monsoon rains, and the trade winds. It is common to find farmers who engage in fishing or collecting fishery resources in mangroves, wine production, trading, and labour services, sequentially or simultaneously, as the season permits. Some farmers are members of the Alitas Farmers Association (AFA), which was organised in 2008. This people's organisation serves as a partner to government in the management of mangroves and nipa palm (*Nypa fruticans*) plantations. This organisation is popular locally for its production of nipa wine, known as *lambanog*. Mangrove cover in Alitas is estimated to be around 355 ha (Fig. 11.3), an area which the AFA has helped restore.

11.2.2 Data Collection

A participatory research approach was adopted for collecting essential socio-demographic, institutional and biophysical information. Data collection activities were conducted from January to mid-March 2020 and involved the active participation of local stakeholders, particularly AFA members, in conducting field data collection activities. The following activities were conducted:

- *Project orientation on the importance of mangrove ecosystems and concept of climate change*
  This initial step involved identification and research coordination with local stakeholders, which included local communities, government institutions (local

Fig. 11.3 Mangrove forest of Alitas, Infanta, Quezon, Philippines (Source: PSA and PhilGIS; Credit: DT Gevaña and JE Garcia)

government units, or LGUs; and the Department of Environment and Natural Resources or DENR), and other relevant stakeholders of the mangrove forests.

- *Collection and analysis of secondary data*

  Secondary information on the study site's bio-physical, socio-demographic, and institutional conditions was collected from various sources including barangay and municipal-level LGUs and DENR.

- *Stakeholder analysis*

  Analysis was conducted through focus group discussions that involved representatives from relevant government agencies and selected members of AFA. These included: (1) an LGU group with six participants; (2) an AFA-member group with ten participants; and (3) a non-AFA-member group with six participants. In addition, key informant interviews with five selected government officials and an AFA officer were also performed to enrich the information obtained from FGDs. Key informant interviews focused on the capacity of stakeholders to carry out adaptation responses, as well as assessment of perceptions on how mangroves are being managed.

- *Resource and climate hazard mapping*

  This involved mapping of natural resources in the community, as well as identification of areas where impacts of climate variability have been observed. The latest satellite images were printed to illustrate the physical condition of the site. Participants were asked to locate: (1) mangrove areas, (2) settlements, and (3) climatic hazards that affected the local community.

- *Institutional capacity analysis*

  This analysis was done through key informant interviews with municipal-level LGU staff (particularly offices on agriculture, environment, local planning and development, disaster risk reduction and management, and social work). A questionnaire was utilised for the interviews, covering seven criteria for evaluating institutional capacity, namely: access rights and entitlements; decision-making processes; information flows; application of new knowledge; capacity to anticipate risk; capacity to respond; and capacity to recover and change. Each criterion was evaluated using a Likert scoring scale: (1) very poor; (2) poor; (3) moderate; (4) good; and (5) very good.

11.3 Key Findings

11.3.1 Climatic Threats

The Alitas community endures numerous natural disasters every year, such as typhoons and tropical storms. A *typhoon* is a tropical cyclone that has a maximum wind speed of 118–220 kilometres per hour (kph) while a *tropical storm* is a tropical cyclone with a maximum wind speed of 62–117 kph. Between 2005 and the time of this study, a total of 21 typhoons and 16 tropical storms visited Alitas.

As recalled by the respondents, the community experienced the following events: Tropical Depression Winnie (2004); Typhoon Tisoy (2019); and a tornado (*buhawi*) (1995) (see Table 11.1). Among these, Tropical Depression Winnie was mentioned as the most destructive. The respondents' recall was very vivid as they shared during the focus group discussions (FGDs) what their life was like during that time. The severity of impacts to locals prompted them to use a rarely used term *delubyo* (a Filipino word for a large flood disaster), to refer to the continuous pouring of rain that caused flash floods and submerged the entire Municipality of Infanta. Fortunately, only 20 houses were flooded, but the water easily subsided. There was no recollection among respondents of significant negative impacts of Typhoon Tisoy. Lastly, only a few locals experienced the tornado, *or buhawi*, (1995) that lasted for an hour and left four houses with damaged roofs.

Along with disasters come risks. As explained by CAPRA (2012), it is certain that the occurrence of a natural, technological, or natural/societal event on a highly vulnerable population will result in human, infrastructure, economic, or financial loss. Risk comprises several elements: primary hazards, exposed elements, and the vulnerability of the community to the event (e.g. poorly-built housing, riverbank construction, and lack of social safety nets). CAPRA is a probabilistic risk assessment platform that aims to strengthen the institutional capacity for assessing, understanding and communicating disaster risk, with the ultimate goal of integrating disaster risk information into development policies and programs.

The risk events experienced by the community made an impact on agriculture and mangroves. Respondents were asked about the degree or extent of the impacts of the risk events (Table 11.2). For Tropical Depression Winnie that resulted in flooding, the FGD participants gave a score of 2 (affected) and explained how agriculture, mangroves, and the environment were affected. Rice fields were greatly damaged by floods and erosion, and mangrove vegetation was removed. Some facilities in the community were also not spared, resulting in a month-long blackout, and potable water supply was lost because several wells were filled with mud. Only those who lived along the river were spared from lack of potable water. Tornado damage was

Table 11.2 Perceived impacts of risk events in Alitas

Risk event	History	Likelihood	Impacts	Degree of impacts (1—not affected 2—affected 3—extremely affected)
Buhawi (Tornado)	1995	Once	Only four houses affected—roofs destroyed	1
Tropical depression, *delubyo* type	Winnie (29 Nov. to 3 Dec. 2004)	First time to experience	Agriculture • Rice fields filled with mud and crops severely damaged Mangrove • Mangrove trees stripped of branches • Some fishes in aquaculture ponds died due to heavy siltation Facilities • Blackout for a month for the entire municipality • Houses along the river damaged • No potable water because a number of wells became silted with mud	2
Typhoon	Tisoy (Dec 2019)	Often	Not affected Barangay Alitas has irrigation canals that helped drain flood from other *barangays*. With the presence of mangroves and nipa, storm surge impact was also lessened.	2

generally rated as 1 (not affected), owing to the mangroves which provided wind breaks.

11.3.2 Mangrove Stakeholders

The direct users of mangrove resources are the local farmers and fisherfolk, particularly AFA members. AFA has 95 members, the majority of which are women (66). One of the programmes of AFA is a mangrove rehabilitation project. This programme was started in 2009 and aimed at reverting abandoned fishponds back to mangroves. In 2013, AFA's efforts were awarded, recognised as a best practice by the Philippine government's National Greening Program. Today, AFA maintains and monitors the mangrove rehabilitation site and plans to expand their rehabilitation efforts to remaining abandoned fishponds.

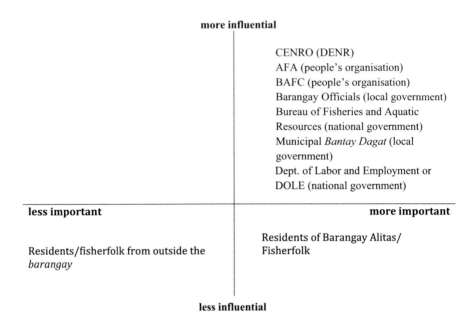

Fig. 11.4 The degree of importance and influence of mangrove stakeholders as perceived by AFA members

The DENR, through its local office called the Community Environment and Natural Resources Office (CENRO), is also an important government institution that assists AFA in its mangrove conservation work. CENRO provides funds for mangrove rehabilitation by local communities and tenure arrangement through the Community Based Forest Management Agreement (CBFMA) programme. This stewardship agreement programme provides AFA with 25 years of stewardship rights (renewable for another 25 years) to utilise and conserve mangroves.

Municipal and barangay LGUs also assist AFA and CENRO in mangrove management. The *barangay*, or community-level LGU, provides the needed ordinances for coastal protection. Together with the municipal-level LGU, it also deputises *Bantay Dagat*, or sea wardens, to protect coastal areas from illegal fishing and timber poaching activities.

Another organisation that influences the management of the mangrove ecosystem is the Barangay Agriculture and Fisheries Council (BAFC). This council is comprised mainly of rice farmers who may or may not be members of AFA. During its meetings in the *barangay*, the farmer members are reminded of the need to protect mangroves for their ecosystem services.

The different user groups and mediating institutions in Alitas are arranged in Figs. 11.4 and 11.5. The diagrams show their relative importance and influence as perceived by members and non-members of AFA, respectively. In these matrices, "importance" refers to the stake of the interest group. The more important the stake is, the more it needs to be considered in policy and management. On the other hand,

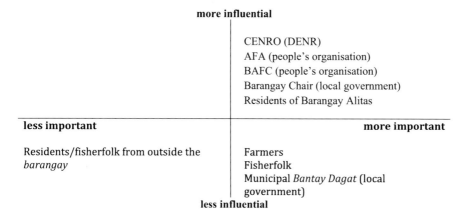

Fig. 11.5 The degree of importance and influence of mangrove stakeholders as perceived by non-AFA members (Credit: Gevaña et al. 2019)

"influence" refers to the capacity to create an impact or influence decisions. For instance, the more influential the interest group is, the more it could cause change in policy and management.

Results showed that there are varying interests and roles involved in mangrove resources and their management. Both AFA members and non-members consider CENRO, BAFC, AFA, and *barangay* LGU officials as highly influential stakeholders with important interests in mangrove rehabilitation as well. Mediating institutions such as the CENRO and the municipal-level LGU were viewed to have vital roles in assisting AFA in conserving mangroves. Both groups (AFA members, and non-members) noted that the interests of the people from outside Alitas concerning the mangrove ecosystem are less important than the rest of the stakeholders. They are also described as less influential in mangrove management.

Overall, mangrove conservation appears as the most common interest and goal among major stakeholders. This implies that forest rehabilitation and protection activities serve as platforms for collaboration among stakeholders.

11.3.3 Socio-Ecological Production System Promoting Local Community Resilience

According to UNU-IAS, Biodiversity International, IGES, and UNDP (2014), resilient socio-ecological production landscapes and seascapes (SEPLS) can be defined as robust systems that demonstrate a balance in the use of natural resources and protection of biodiversity in promoting human well-being. This concept adheres to the three socio-ecological tenets of: (1) consolidation of wisdom on securing diverse ecosystem services; (2) integration of traditional and ecological knowledge with scientific knowledge; and (3) adoption of collaborative management. The

Fig. 11.6 Mangrove-dependent livelihoods of Alitas community (Credit: DT Gevaña and E O'Brien)

resilience of the SEPLS in Alitas is best exemplified by the local livelihoods that mangrove conservation supports, local knowledge, and stakeholders' commitment to work together in protecting mangroves. Community resilience is commensurate of the mangrove areas that are well-kept.

Rehabilitation and protection of mangroves (locally termed *pakatan*) and nipa zones (locally termed *nipahan*) are collective goals aimed at keeping the community adaptive to climate hazards. Different user groups benefit from the rehabilitated and protected mangrove areas either directly or indirectly. The direct user groups of the *pakatan* are the residents of Barangay Alitas (Fig. 11.6). The fisherfolks benefit directly from three main types of marine resources extracted from the mangrove ecosystem: fish, shrimp and shellfish. Aside from fishing, other livelihoods include production of *lambanog* (nipa wine) and nipa shingles, sold as roofing material. There are three *lambanog* distilleries owned by AFA, and many other privately operated distilleries located in Alitas, where residents process nipa into *lambanog*.

AFA members also obtain additional income from joining mangrove rehabilitation projects of the DENR (Fig. 11.7). They provide labour in nursery production, field planting, and site maintenance activities (e.g. replanting and brushing). Women dominate in nursery seedling production activities, while men are engaged more in field planting and plantation maintenance.

Local residents regard the *pakatan* to be essential in protecting them from typhoons, strong winds, and monsoon rains, serving as a buffer zone that minimises the risk of flooding. Residents also collect fallen branches and driftwood called *daghip* from the *pakatan* to use as fuelwood and fence material for their plant nurseries. *Daghip* are trunks or branches of trees damaged by strong winds that drift ashore. Figure 11.8 shows a map of land use and climatic hazards that was generated through participatory mapping.

Tapok—concrete houses serving as evacuation areas for vulnerable families during strong typhoons

Fig. 11.7 Mangrove rehabilitation project as an additional local income source (Credit: DT Gevaña)

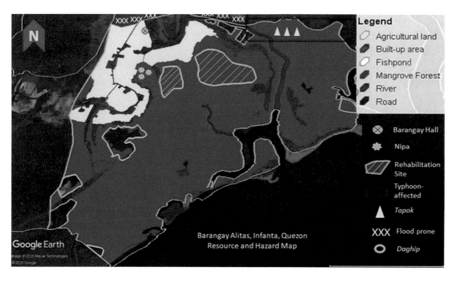

Fig. 11.8 Community resource and hazard map of Alitas community (Source: Google Earth ver. 1.2.2019; Credit: JE Garcia)

Daghip—driftwood and damaged branches collected as fuelwood and fence material for nurseries

The mangrove ecosystem contributes to local resiliency. People living within it and those from nearby *barangay* can freely fish or glean for shellfish in mangrove areas for food and income. Local people also have high regard for the important role of mangroves in providing nursery grounds for marine fishes, particularly the *tulingan* (tuna), which is a major product of the town. Every year, the whole town of Infanta celebrates the *Tulingan* Festival to commemorate abundant blessings from the sea and mangroves. Further, policies prohibiting the cutting of mangroves and the use of destructive fishing methods such as *salambaw* (poisoning) and *lumpot* (use of fine nets with less than 3 cm diameter mesh) are strictly enforced.

Local resilience is also reflective of effective local climate change adaptation strategies that rest on maintaining healthy mangrove cover. During strong typhoons, some residents evacuate to *saguban*, a temporary shelter made of nipa that is located inside the nipa plantation and mangroves.

11.3.4   Institutional Capacity

Table 11.3 summarises findings on the institutional capacity of the municipal-level LGU to promote local resilience and the sustainable management of mangroves. In terms of *application of new knowledge* in addressing climatic hazards and conserving mangrove resources, the key informant interview respondents regarded the LGU's capacity to be 'very good', with a mean score of 4.8. The level of awareness and knowledge on climate change and the mangrove ecosystem of Alitas was also observed to be very good.

In view of *access rights and entitlements*, including the human, physical and financial resources of the LGU, respondents indicated that the agency has good capacity, with a score 4.0. For human resources, plans and programmes exist to increase the number of staff (with permanent positions and ample trainings) to ensure that the LGU's mandates for enhancing resiliency and protecting the environment are well implemented. In terms of financial resources, respondents underscored a certain degree of need to improve fund allocation for disaster management and mangrove management. Further, sourcing of funds from private institutions was mentioned as a good strategy to address budget deficits.

Decision-making processes for disaster management and mangrove conservation were described to be generally effective. Respondents noted that there are concrete and clear protocols (on administrative, legislative and executive mandates) that guide offices and staff to develop and implement programmes. Further, the conduct of regular meetings was also regarded as helpful in ensuring timely and coordinated

Table 11.3 Summary of scores on the institutional capacity of LGU-Infanta for climate-related risk management

Institutional capacity/resilience indicators	Median	Description of institutional capacity
Access rights and entitlements	4.0	Good
Decision-making processes	4.0	Good
Information flow	3.5	Moderate to good
Application of new knowledge	4.8	Good to very good
Capacity to anticipate risk	4.5	Good to very good
Capacity to respond	4.0	Good
Capacity to recover and change	4.0	Good

completion of tasks and targets. With all the protocols in place, flexibility to change decisions or correct actions (as needs arise) was, however, identified as an area needing improvement.

The LGU's institutional *capacity to anticipate risk* (4.5), *capacity to respond* to risk (4.0), and *capacity to recover and change* during disasters (4.5) were all perceived to be 'good'. This capacity is evident in the strict observance of protocols on disaster risk management, as mandated by national law and monitored by the national government.

In terms of *information flow*, which is defined to as how effectively other stakeholders (such as the local community of Alitas and private organisations) can easily access accurate information from the LGU about risk management and climate change adaptation, respondents found the LGU's capacity to be 'moderate'. Improving the LGU's capacity in data management information systems was mentioned as necessary to be able to package and communicate plans and programmes on disaster and forest management to local communities and the broader public. Furthermore, skills in processing essential data (e.g. meteorological information and geo-hazards maps) must be improved in order to formulate correct and updated plans.

11.3.5 Collaborative Mangrove Conservation Promoting Local Resilience

Figure 11.9 provides a conceptual framework and synthesis of the perceived importance of conserving mangroves in promoting local development, also depicting how climate resilience can be enhanced through improvement of the government's institutional capacity. This framework also underscores collaborative management among major stakeholders to encourage: (1) local commitment and participation in mangrove protection; and (2) capacity improvement of different institutions in climate and disaster risk management. Lessons from the case of Alitas suggest

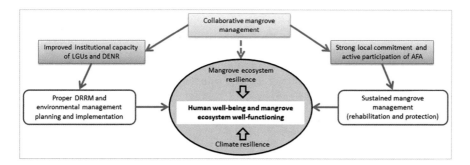

Fig. 11.9 Conceptual framework on how mangrove conservation promotes local resilience in Alitas (Credit: DT. Gevaña et al. 2019)

how investing in mangrove conservation reverberates to achievement of well-being for humans and a well-functioning ecosystem.

Furthermore, the distinct link between mangrove conservation and enhanced local resilience depicts the real essence of socio-ecological production landscape and seascape (SEPLS) management. Mangroves were generally perceived as the hinge that links various ecosystems (within the landscape) upon which diverse livelihoods are very much dependent. By pursuing collaborative mangrove management, community development is also promoted as: (1) the local community develops a sense of ownership, organisation and responsibility over mangrove management; (2) women and marginalised sectors are empowered through participation in mangrove conservation activities; (3) government and non-government stakeholders take on responsibility to assist the local community in mangrove protection efforts, thereby satisfying their institutional goals as well; (4) mangrove programmes (particularly reforestation) serve as a conduit for creating jobs and local income sources; (5) well-kept mangrove forests sustain and protect natural resource-dependent local livelihoods (e.g. farming, fishing, and nipa wine production) from the injurious impacts of climate change; and (6) new livelihoods (e.g. ecotourism) are developed that promote self-reliance in funding and sustain mangrove conservation activities.

11.4   Discussion: Implications for Transformative Change

Pursuing collaborative mangrove management promotes socio-ecological well-being and the resilience of stakeholders, particularly the local community of Alitas. The community now needs to harness and nurture these seeds of transformative change. These include the following:

- Local stakeholders have learned to regard mangrove conservation as an effective climate change adaptation strategy. They noted that keeping a good mangrove cover will lessen their vulnerability to the impacts of typhoons and floods. Likewise, local residents understood well that mangroves provide diverse ecosystem services, and these can only be sustained if they are properly managed. Continuously sharing such knowledge and appreciation of the mangrove's value with broader stakeholders and generations to come is essential.
- Active local participation in rehabilitation projects. This was regarded to be beneficial in creating additional income sources for the local community. Likewise, the intimate involvement of local people, particularly the AFA members, has also strengthened their sense of stewardship and commitment towards protecting the mangroves they planted.
- Local government units and CENRO are committed to assisting the local community in mangrove conservation. Such commitment is expressed in terms of the provision of livelihood projects, environmental law enforcement, and policy improvements that complement the community's interest in conserving

mangroves. This sense of constituent governance enhances the local people's trust and support for the government, while assuring the achievement of its mandates and goals to promote local development as well.

- CBFMA provides a legal and long-term platform for sustaining mangrove rehabilitation efforts by encouraging more stakeholder collaboration. This tenure programme provides clear guidelines on the rights, roles and responsibilities of the AFA to serve as stewards and a partner of the government in mangrove conservation. It also highlights that the centrality of mangrove management rests on them. This opens up opportunities for other stakeholders (particularly private institutions and NGOs) to partner with AFA in local development projects that benefit the local community and the mangroves they manage.

11.5 Conclusion and Recommendations

This paper looked at the importance of managing mangroves to address climate change hazards and promote the sustainability of a socio-ecological production landscape and seascape. It synthesised various perspectives on the importance of mangrove conservation for enhancing climate resiliency by: (1) describing the climate-related hazards that affect local communities and mangroves; (2) describing the socio-institutional structures influencing mangrove conservation and local resilience; and (3) identifying climate change adaptation strategies that promote local development and mangrove conservation.

Climatic hazards, as a result of tropical depression was identified to have the most severe negative impact on the local community. Keeping a good cover of mangrove was recognised as key for counteracting severe impacts. Furthermore, mangrove conservation was perceived to have sustained various local livelihoods such as nipa wine production, farming and fishing. Ensuring the continuous delivery of these protective and provisioning services necessitates socio-institutional transformations that are pegged on a collaborative management platform. Findings from stakeholder analysis and assessment of institutional capacities on disaster risk management yielded two important lessons. First, collaborative management can be realised by harmonising diverse stakeholders (in terms of their interests) to help the local community (particularly AFA) manage their mangroves well. Second, the centrality of promoting climate resiliency rests on the ability and willingness of various stakeholders to work together.

There are evident seeds of transformative change that need to be nurtured. First, local stakeholders regard mangrove conservation as an effective climate change adaptation strategy. Second, there is active local participation in rehabilitation projects. Local commitment to rehabilitation is sustained by the diverse benefits that the local community enjoys from engaging in it. Third, government stakeholders are committed to assisting the local community in mangrove conservation. Lastly, adoption of the CBFMA tenure programme can provide a legal and long-term platform for sustaining mangrove rehabilitation efforts by encouraging more stakeholder collaboration with AFA and DENR.

Overall, this study underscores collaborative management as a prerequisite for a sustainable SEPLS and climate resiliency in the mangrove landscape. The best practices shown herein and the seeds of transformative change that have stimulated multi-stakeholder mangrove conservation could provide some lessons which other CBFMA programmes in the country could replicate or learn from.

Acknowledgements This case study was made possible through the funding support of the Institute for Global Environmental Strategies (IGES) through the Satoyama Development Mechanism (SDM) Project; and through the warm accommodation and active participation of Barangay Alitas, particularly AFA members, LGU of Infanta, and DENR-CENRO Real, Quezon during the conduct of field data collection.

References

Akter, S. (2015). Climate change hazards vulnerability and resilience capacity assessment for Char land women in Bangladesh. PhD Thesis, University of Dhaka, Bangladesh.

Ali, F., Ingirige, B., & Abidin, N. (2018). Assembling and (re)assembling critical infrastructure resilience in Khulna City, Bangladesh. *Procedia Engineering, 212*, 832–839.

Abarquez, I., & Murshed, Z. (2004). *CBDRM field practitioners' handbook*. Bangkok: Asian Disaster Preparedness Center (ADPC).

Center for Community Enterprise. (2000). The community resilience manual: A resource for rural recovery and renewal, The Center for Community Enterprise, British Columbia, Canada.

Chua, T. (1998). Lessons learned from practicing integrated coastal management in Southeast Asia. *Ambio, 27*, 599–610.

Camacho, L., Gevaña, D., Carandang, A., Camacho, S., Combalicer, E., Rebugio, L., & Youn, Y. (2011). Tree biomass and carbon stock of a community-managed mangrove forest in Bohol, Philippines. *Forest Science and Technology, 7*(4), 161–167.

CAPRA. (2012). *The important distinction between disaster and risk*. Retrieved 14 March, 2020, from https://ecapra.org/documents/important-distinction-between-disaster-and-risk

Frankenberger, T., Mueller, M., Spangler, T., & Alexander, S. (2013). *Community resilience: Conceptual framework and measurement feed the future learning agenda*. Rockville: Westat.

Garcia, K., Malabrigo, P., & Gevaña, D. (2014). Philippines' mangrove ecosystem: Status, threats and conservation. In I. Faridah-Hanum, A. Latiff, K. R. Hakeem, & M. Özturk (Eds.), *Mangrove ecosystems of Asia: Status, challenges and management strategies* (pp. 81–94). New York: Springer.

Gevaña, D., & Im, S. (2016). Allometric models for Rhizophora stylosa Griff. in dense monoculture plantation in the Philippines. *Malaysian Forester, 79*(1 & 2), 39–53.

Gevaña, D., Camacho, L., & Pulhin, J. (2018). Conserving mangroves for their blue carbon: Insights and prospects for community-based mangrove management in Southeast Asia. In C. Makowski & C. W. Finkl (Eds.), *Threats to mangrove forests* (Vol. 25, pp. 579–588). New York: Springer.

Gevaña, D., Pulhin, J., & Tapia, M. (2019). Fostering climate change mitigation through a community-based approach: Carbon stock potential of community-managed mangroves in the Philippines. In R. R. Krishnamurthy, M. P. Jonathon, S. Srinivasalu, & B. Glaeser (Eds.), *Coastal management: Global challenges and innovations* (pp. 271–282). London: Elsevier.

Honda, K., Nakamura, Y., Nakaoka, M., Uy, W., & Fortes, M. (2013). Habitat use by fishes in coral reefs, seagrass beds and mangrove habitats in the Philippines. *PLoS ONE, 8*(8), 1–10.

IPCC. (2012). Glossary of terms. In C. Field, V. Barros, T. Stocker, & Q. Dahe (Eds.), *Managing the risks of extreme events and disasters to advance climate change adaptation: A special report of the Intergovernmental Panel on Climate Change* (pp. 555–564). Cambridge: Cambridge University Press.

Mallick, F. (2013). Habitat and infrastructures: A localized approach to resilience. In R. Shaw, F. Mallick, & A. Islam (Eds.), *Climate change adaptation actions in Bangladesh* (pp. 331–340). Tokyo: Springer.

Menéndez, P., Losada, I., Torres-Ortega, S., Toimil, A., & Beck, M. (2019). Assessing the effects of using high-quality data and high-resolution models in valuing flood protection services of mangroves. *PLoS ONE, 14*(8), 1–14.

Moles, O., Caimi, A., Islam, M., Hossain, T., & Podder, R. (2014). From local building practices to vulnerability reduction: building resilience through existing resources, knowledge and know-how. *Procedia Economics and Finance, 18*, 932–939.

Mustelin, J., & Handmer J. (2013) Triggering transformation: Managing resilience or invoking real change? in *Proceedings of transformation in a changing climate conference*, University of Oslo, pp. 24–32.

Ostadtaghizadeh, A., Ardalan, A., Paton, D., Jabbari, H., & Khankeh, H. R. (2015). Community disaster resilience: A systematic review on assessment models and tools. *PLOS Currents Disasters, 7*, 1–16.

Pal, U., Bahadur, A., McConnell, J., Vaze, P., Kumar, P., & Acharya S. (2019). Unpacking transformation: A framework and insights from adaptation mainstreaming. ACT learning paper, action on climate today, Oxford Policy Management, UK.

Pfefferbaum, B., Pfefferbaum, R., & Van Horn, R. (2014). Community resilience interventions: Participatory, assessment-based, action-oriented processes. *American Behavioral Scientist, 59*, 238–253.

PSA. (2018). Census of population 2018, Philippine Statistics Authority (PSA), Manila, Philippines.

Renaud, F., Sudmeier-Rieux, K., & Estrella, M. (2013). *The role of ecosystems in disaster risk reduction* (p. 486). Tokyo: United Nations University Press.

Sameen, S. (2018). Process inclusive infrastructure: notions towards cyclone resilience in Bangladesh. *Procedia Engineering, 212*, 30–38.

Sharifi, A., & Yamagata, Y. (2016). On the suitability of assessment tools for guiding communities towards disaster resilience. *International Journal of Disaster Risk Reduction, 18*, 115–124. https://doi.org/10.1016/j.ijdrr.2016.06.006.

Sharifi, A. (2016). A critical review of selected tools for assessing community resilience. *Ecological Indicators, 69*, 629–647.

Turnbull, M., Sterrett, C., & Hilleboe, A. (2013). *Toward resilience: A guide to disaster risk reduction and climate change adaptation*. Warwickshire: Practical Action Publishing Ltd.

UNISDR. (2007). *Hyogo framework for action 2005–2015: Building resilience of nations and communities to disasters*. Geneva: United Nations International Strategy for Disaster Reduction.

UNU-IAS, Biodiversity International, IGES & UNDP. (2014). Toolkit for the indicators of resilience in socio-ecological production landscapes and seascapes (SEPLS).

US Indian Ocean Tsunami Warning System Program. (2007). *How resilient is your coastal community? A guide for evaluating coastal community resilience to Tsunamis and other coastal hazards*. Bangkok: US Indian Ocean Tsunami Warning System Program supported by the United States Agency for International Development.

White, A. T., Christie, P., D'Agnes, H., Lowry, K., & Milne, N. (2005). Designing ICM projects for sustainability: Lessons from the Philippines and Indonesia. *Ocean and Coastal Management, 48*, 271–296.

The opinions expressed in this chapter are those of the author(s) and do not necessarily reflect the views of UNU-IAS, its Board of Directors, or the countries they represent.

Chapter 12
Improvement of Human and Environmental Health Through Waste Management in Antigua and Barbuda

Ruth Viola Spencer

Abstract Antigua and Barbuda is currently experiencing an expansion in integrated waste management driven by local community groups. These events are catalytic and transformational, fit well into SEPLS methodologies, and contribute to many of the biodiversity targets and Sustainable Development Goals (SDGs). With waste being a direct driver of and major challenge for biodiversity, climate change and land degradation, many positive multi-dimensional impacts are being seen, realised and manifested that contribute positively to reducing land-based sources of pollution through community stewardship. Such local actions positively impact the sustainable management of natural resources and the protection of habitats. Likewise, they support land degradation neutrality, protection and safeguarding of the ecosystems that provide our soil, air and water resources, sustain livelihoods, and facilitate transfer of knowledge to children and youth.

This attempt to pilot a circular economy approach is providing vocational and life skills training, as well as income generation for the local community, including new forms of capacity building and development, while reducing soil, water and air pollution. Public-private partnerships built through this project are motivating other groups to follow a similar path to biodiversity transformation. Changes in attitudes and behaviours, and the building of knowledge and capacities in the next generation, is taking place through school and community outreach programmes.

The project has led to community empowerment in understanding that everyone has a role to play in sustainable development and that through collective actions, changes to improve public health can be made.

Keywords Commitment · Actions · Partnerships · Changes in behaviours and attitudes · Waste management

R. V. Spencer (✉)
Marine Ecosystems Protected Areas (MEPA) Trust, St Johns, Antigua and Barbuda
e-mail: ruth.spencer@mepatrustantiguabarbuda.org

© The Author(s) 2021
M. Nishi et al. (eds.), *Fostering Transformative Change for Sustainability in the Context of Socio-Ecological Production Landscapes and Seascapes (SEPLS)*,
https://doi.org/10.1007/978-981-33-6761-6_12

12.1 Introduction: Context and Challenges

Antigua and Barbuda is a Caribbean Small Island Developing State (SIDS) and recognised as a biodiversity hotspot with many fragile and sensitive ecosystems, including dry forests, wetlands, salt ponds and oceans (Antigua and Barbuda Meteorological Service 2020). The islands are challenged by natural hazards including hurricanes, earthquakes, fires, floods, and in recent years, excessive influx of sargassum seaweed. Residents are very dependent on the oceans for food and livelihoods, leading to vulnerability towards the impacts of climate change (Table 12.1).

Our wetlands include a variety of ecosystems, such as mangroves, lagoons, estuaries, and marshes, which host endemic species, as well as provide ecosystem services including water drainage and filtration, food resources and coastal stabilisation, enhancing resilience against natural hazards. Oceans provide ecosystem services including tourism, which generates one of the largest financial incomes for the country; however, they are impacted by overfishing, land-based pollution, ocean warming, sea-level rise and invasive alien species.

Antigua has signed on to most international conventions and has many legislations on the books, including the Environmental Protection and Management Act (No. 10 of 2019) which established an integrated and sustainable environmental management system. However, enforcement remains an issue. Key for buy-in and local ownership is making the linkages between biodiversity conservation or environmental health and human health, since we all live in the same environment. The threats that unsustainable development wield directly on our fragile ecosystems undermine the important role these ecosystems play in disaster risk reduction and providing services and functions to people and nature. Accordingly, this role needs to be better understood by policymakers. Land use needs to be better planned, such as by integrating mandatory environmental impact assessments. Unplanned land clearing and development for housing and tourism is leading to poor and uncovered

Table 12.1 Basic information of the study area

Country	Antigua and Barbuda
Province	Saint John Parish
District	Cooks
Municipality	St. John's City and St. John's Rural
Size of geographical area (hectare)	1000 (St. John's City)
Number of indirect beneficiaries	21,643 (population of St. John's City)
Dominant ethnicity(ies), if appropriate	West African, British, and Madeiran descent
Size of case study/project area (hectare)	1.56 (Cooks Landfill[a])
Number of direct beneficiaries	600
Dominant ethnicity in the project area	African descent
Geographic coordinates (latitude, longitude)	17° 7′ 11.28″ N; 61° 51′ 27.00″ W

[a]The operational surface area of Cooks Landfill is reported at 15,600 m^2 (Francis et al. 2015)

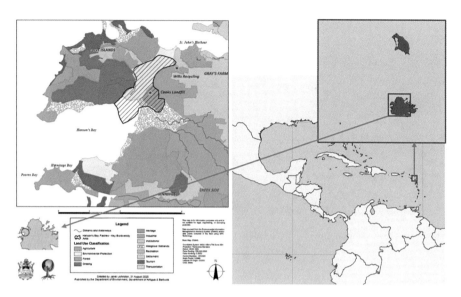

Fig. 12.1 Map of the study area

soils. When it rains, the soil erodes, not only generating risk of landslides, but also degrading the water quality of rivers and oceans.

The study area is located in the Cooks District of St John's City, the capital of Antigua and Barbuda. Over 20% of the national population resides in St John's City, and it is also the gateway for over one million tourists to the nation every year. A majority (73%) of tourists arrive on cruise ships (Antigua and Barbuda Tourism Authority 2019). With a rapidly increasing tourist population, waste management has been a concern in the study area. A mechanism for collecting waste from cruise ships exists in accordance with Annex V of the International Convention for Prevention of Pollution from Ships (MARPOL 73/78), however, it has been pointed out that the system needs improvement to achieve sustainability. All waste from the nation is officially disposed of at one site, the Cooks Landfill (Gore-Francis and Ministry of Agriculture, Housing, Lands and the Environment 2013). Household waste received at the Cooks Landfill as of 2014 was 20,909 t/year (Francis et al. 2015). The Cooks Landfill site is situated next to mangrove wetlands and Hanson's Bay Flashes (Fig. 12.1), a designated International Birding Area and feeding ground for the Vulnerable West Indian Whistling-duck (*Dendrocygna arborea*) (BirdLife International 2020). Hanson's Bay Flashes consists of a terrestrial ecosystem (a low-lying mangrove wetland) and a marine ecosystem (mostly undeveloped bay with seagrass and coral reef). It also serves a critical function as a watershed, and is deemed an area of hydrologic importance (Devine et al. 2010).

Normal annual rainfall between 1981 and 2010 was 1049.2 mm. Antigua has been facing severe drought conditions since March 2020 (Antigua and Barbuda Meteorological Service 2020).

Fig. 12.2 Cooks landfill (source: Marlon Jeffers, Belmont Studios)

Drought conditions exacerbate water issues, and the single utility company covering the entire islands of Antigua and Barbuda spends considerable amounts of resources to purify polluted groundwater supplies of chemical substances, including those which may be leaching out from the Cooks Landfill (Fig. 12.2). The landfill currently lacks testing facilities and monitoring points. The local community does not have access to scientific data or measurements of the impacts of the pollution, but can visibly see the massive degradation of the surrounding mangroves and other vegetation. Currently domestic waste is not segregated when it is collected, and all kinds of waste are carried to the landfill. Sanitary waste is stored at an old disposal site.

To improve the waste management in the country, education and awareness-raising is carried out at the community level, such as through schools and church programmes. Through education and awareness-raising, it is hoped that enforcement of rules and legislation, such as the National Solid Waste Management Authority Act, will improve.

12.2   Local Actions

The local actions of Wills Recycling (WR), a private enterprise, and the Zero Waste Antigua Barbuda (ZWAB) show proactive work that the outcomes of the 2019 Rotterdam negotiations can build on for several waste issues, including marine plastic litter and microplastics (Fig. 12.3). WR/ZWAB has been building synergies and partnership with local groups, including the Antigua Barbuda Waste Recycling Company (ABWREC). The West Indies Sailing Heritage (WISH) Foundation's Ocean of Opportunities Project for Plastic Upcycling is determined to make a national breakthrough in pioneering the full utilisation of hundreds of tons of

Fig. 12.3 ZWAB meeting (Source: ZWAB)

Fig. 12.4 Wills recycling (source: Marlon Jeffers, Belmont Studios)

shredded plastic waste lying idle in Antigua to make furniture. Funds for designing the technology are currently being sought for this process.

WR felt the compulsion to take action after observing daily the number of truckloads of mixed waste material going into the Cooks Landfill, only to be buried by heavy machinery. WR's visionary and pioneering effort led to the set-up of a metal waste recycling centre right outside of the entrance to Cooks Landfill, and each day it is abuzz with activity (Fig. 12.4). WR's concern for human and environmental health has been translated from ideas into actions, including support for waste pickers, many of them women, using their bare and exposed hands to dig into the heaps of garbage. WR provided advice, training and a cash incentive to enable them to have an income. WR also saw oil and dangerous metal residues leaking into the soils and impacting the wetlands located next to the landfill, anticipating high risks to soil and ground water.

In the context of transformative change as proposed by IPBES (2019), the project conducted by WR strongly influenced the following levers: incentives and capacity-building; cross-sectoral co-operation; and decision-making in the context of resilience and uncertainty. The leverage points shown to be effective in the context of this project were: visions of a good life; reduction of waste, values and action; inequalities; justice and inclusion in conservation; technology, innovation and investment; and education and knowledge generation and sharing. In addition, the impacts from the direct and indirect drivers of environmental degradation were significantly reduced through the strengthening of socio-ecological resilience systems.

12.3 Results

Only 3% of global environmental funding is spent to address chemical and waste issues. Yet, in Antigua and Barbuda, local actions propelled by passion, commitment, and dedication are robust, strong and directed to pioneering good practices for waste management. The focus on the concept of one health for humans and the environment is attracting youths. People have experienced a striking realisation that health depends on clean air, soil and water, and that positive collective action is required because the issue affects everyone.

The relationships between key stakeholders in this project are depicted in Fig. 12.5. WR also provides training in the protection of personal health required for the waste sorting process. The ongoing building of relationships with key private sector groups and state agencies is having a catalysing impact and motivating other groups to follow the path to biodiversity transformation in their villages and communities. Relationships are being built and strengthened and conversations are ongoing, even with other scrap metal dealers as the granulator set up by WR on

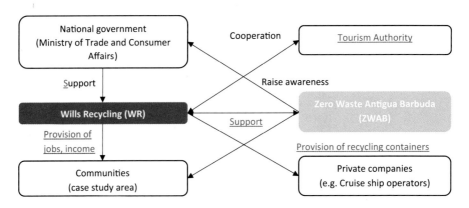

Fig. 12.5 Key project stakeholders (Source: author)

the island is being positioned to stop all open burning. The buy-in of other stake-holders is encouraging and is leading to positive impacts on human and environ-mental health.

The transformation process involves efforts to change attitudes and behaviours by building the knowledge and capacities of boys, girls, men and women, who through their involvement in the school and community outreach programmes and visits to observe the actions at the WR waste site, are able to learn and see how waste, which previously was buried with harmful effects on our health and our environment, can be repurposed and reused. The positive outcomes on people's health are evident, especially the health and well-being the of the residents in the nearby densely populated districts of Green Bay, Gray's Farm, Perry Bay and Hatton.

WR's efforts at building coalitions with cruise ships, hotels and utility companies to redirect their waste streams supported by related inter-agency dialogue and national sharing processes, is also providing content for national public education, as well as outreach and advocacy on the waste challenges facing the country and local workable solutions. Building synergies demonstrates a unified approach that brings high visibility at the national level and a pathway to national buy-in, which in turn brings about changes in the attitudes and behaviours of the people. Teachers are joining the local groups and advocating for the changes through school systems and educational processes. WR's efforts have been showcased on national media, fea-tured in six national reports and a video made of its efforts to safeguard biodiversity. WR is one of the key groups selected for the International Union for Conservation of Nature (IUCN) "Plastic Waste-Free Islands" project, working with Asia Pacific Waste Consultants to engage with local people through surveys and provide data on plastics and microplastics.

WR has been manufacturing movable waste storage containers recycled from waste iron and other metals that hotels, cruise ships and utility companies can utilise in the sorting and separation process of their waste management. Most of the hotels and cruise ship ports are located in marine areas. WR is represented as one member of the executives of the newly formed Zero Waste Antigua Barbuda, an NGO, and has been very active in the "Clean is Cool" campaign in partnership with the Tourism Authority. This campaign carries out strong outreach in villages and schools and is currently implementing a national programme for the phase out and phase down of mercury by 2020. WR has been involved in providing safe disposal for light bulbs and fixtures containing mercury. In activities spearheaded by the Ministry of Trade and Consumer Affairs, WR is a key player in demonstrating actions that residents can emulate to become "sustainable consumers".

The planning of programmes and policies related to this project has always included a wide range of partners with the vision of working together to share knowledge and information based on inclusion and wide participation in a whole-of-society approach. This has been widely recognised in the country (Annex 1).

The challenge is to bring understanding and awareness to everyone through a bottom-up approach, especially to the government technicians who draft policy documents. Efforts towards mainstreaming and integrating information from other stakeholders in other sectors are taking place. The Department of Environment's Technical Assistance Committee must be applauded for setting up an inter-ministerial group which meets monthly and is now going on its 50th meeting. The meetings include a wide range of stakeholders, such as government technicians, the private sector, NGOs and local community representatives.

With the formation in 2018 of Zero Waste Antigua Barbuda (ZWAB), WR began to serve as treasurer and mentor of the group, which has a strong youth following (Fig. 12.6). WR has been actively involved in the ongoing programmes and activities of the new group, using it as a platform for wider stakeholder involvement. These include implementing a national mercury management programme in line with the targets of the Minamata Convention on Mercury. The programme is implemented through various pathways, including involvement of medical clinics and facilities on the island, development of new alternative products, and training of customs officials for awareness and capacity building.

The socio-ecological production landscapes and seascapes (SEPLS) approach helps to eliminate harmful practices, which in the long run negatively impact the health of ecosystems and result in deforestation and land degradation, leading to soil erosion, landslides, and flooding that impact communities. Antigua and Barbuda, as a small country, cannot afford to lose the fundamental services that our ecosystems provide, such as watershed services, soil protection, erosion control, disaster risk reduction, carbon sequestration and climate regulation. Our tourism sector, which provides 75% of our GDP, has also faced impacted with multiple types of liveli-hoods that support many families through eco-tourism activities at risk. Likewise, hikes, walks through nature trails, birding and other recreational services all support human and environmental health and wellness activities.

Forest and other ecosystem restoration contribute to many goals and targets set at the global and regional level under multilateral environmental agreements and other frameworks. The local groups in Antigua and Barbuda have benefited from capacity building, knowledge sharing and networking support targeted at the Caribbean and provided through the support of the Korean Government's Forest Ecosystem Res-toration Initiative (FERI). The knowledge received is helping to support local actions in the implementation of the Strategic Plan for Biodiversity 2011–2020 and the Aichi Biodiversity Targets. This knowledge is also helping in the peer review process for the Global Biodiversity Framework (GBF), and the SEPLS approach has brought many stakeholders to share in discussions and deliberations.

Antigua and Barbuda has committed and set Land Degradation Neutrality (LDN) targets, and WR, a member of ZWAB, which is accredited by the United Nations Convention to Combat Desertification (UNCCD), is also seeking membership in the Global Mercury Platform. WR is actively promoting local actions and building synergies and partnerships with extensive outreach to meet the LDN targets. Some-times it is the proactive local actions of groups like WR and the engagement of private citizens that awaken and spur policymakers and technicians to take actions at the national government level to protect and conserve our ecosystems. Due to overlap between the main drivers of land degradation, soil and biodiversity loss,

Fig. 12.6 "Clean is cool" events with the Tourism Authority to educate and bring awareness to youths (Source: ZWAB)

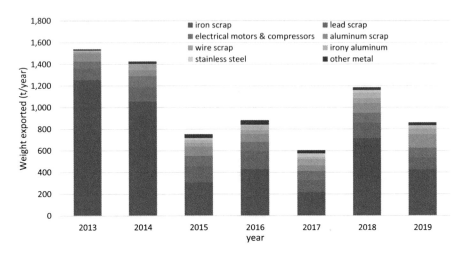

Fig. 12.7 Summary of metal waste exported from Antigua and Barbuda for recycling (2013–2019)

there is great potential to bring LDN targets and policies in line with existing and future commitments on biodiversity and climate change.

WR is one of the first groups to feed data and information on metal waste into the environmental registry being set up in the government-led Environmental Information and Management System (EIMAS). WR has also made an effort to collect and share information that has been fed into the 6th National Report to the CBD. Moreover, new and additional data on adaptation and mitigation actions will also be fed into the Biennial Update Report (BUR) for the United Nations Framework Convention on Climate Change.

Since its inception in 2013, WR has exported more than 7500 tons of metals for recycling (Fig. 12.7). According to WR, cost of goods sold, expenses and profit respectively account for 37%, 48%, and 15% of revenue.

12.4   Findings and Lessons Learned

The concepts of SEPLS fit into many landscapes where humans, nature, local culture, heritage and knowledge can all be linked and fully integrated into local biodiversity. In the case of the local actions taken in Antigua and Barbuda described above, the key considerations of stakeholder engagement and the continual evaluative process, as well as shared understanding and collective visioning, have led to the empowerment and capacity building of local community groups, with Wills Recycling leading and pioneering this process. The steps of gathering knowledge and information, and developing strategies and plans with stakeholders to implement actions, are both realised and ongoing. WR has provided people with awareness of how well-being is dependent on nature and its contributions to people, showing that highlighting the multiple values of ecosystem functions and nature's contributions to people leads to better outcomes. This case study demonstrates a bottom-up approach to transformative change, whereby multi-stakeholders continued to be proactive in finding and supporting community-driven approaches to protect and conserve the environment and to building, strengthening and promoting integrated landscape management. This case study will enable local groups to get recognition and visibility and will help to build new and additional skillsets and capabilities. Resources and technical support for data gathering, scientific studies and other IT upgrades are needed, thus the project seeks to access opportunities that can extend and enlarge our boundaries. WR actions have shown many benefits, including enhancing MEA synergies, alignment with many targets, and promoting knowledge and awareness among youth, children, teachers and private citizens.

Use of the SEPLS approach in a small island context aimed at societal transformative changes is gradually being manifested in results on many development objectives and represents a workable and observable alternative to other conventional efforts. It fully supports a whole-of-society approach that includes the involvement and participation of private sector businesses, with many new partnerships and coalitions being developed. Citizen engagement is also growing. Now, when the local people see harmful practices and actions, they have begun to make these known publicly across the airways on talk shows or social media, bringing attention to the negative impacts on our biodiversity.

Such growing local awareness and actions are contributing positively to the long-term sustainability of our natural resources, protection of habitats, protection and safeguarding of the ecosystems that provide our soil, air and water resources, livelihoods creation, and transfer of knowledge to children and youth. Likewise, supporting and mentoring other community-led biodiversity enhancement efforts is leading to the upscaling of similar programmes.

Challenges remain in gathering scientific evidence for accelerating implementation and monitoring. There is an urgent need for monitoring points to be set up across the island and in the wetlands around the Cooks Landfill to capture data that can be shared in scientific studies and to measure the impacts on human and environmental health of the various forms of pollution. The proactive efforts described herein will continue to be included in calls and proposals directed at governments, particularly related to issues that target local actions and have strong opportunities for partnerships in national projects aimed at building capacities. The way forward calls for inclusive and participatory governance processes.

Annex 1

Newspaper article: "Public urged to participate in 'plastic free' data gathering exercise" (Orville Williams, *The Daily Observer*, 19 February 2020).

Public urged to participate in 'plastic free' data gathering exercise

Members of the Asia Pacific Waste Consultants company (dressed in blue), alongside representatives of the Ministry of Health, Wellness and the Environment (dressed in white) and Zero Waste Antigua and Barbuda (dressed in green).

By Orville Williams

Residents of Antigua and Barbuda are being urged by the Ministry of Health, Wellness and the Environment, to participate in a data gathering exercise taking place over the next couple of weeks.

The project – called the Plastic Free Waste Island project – will run for three years and is being led by the International Union for Conservation of Nature, Regional Office for Mexico, Central America and the Caribbean (IUCN-ORMACC), with support from the Norwegian Agency for Development Cooperation (NORAD), through the Ministry of Health, Wellness and the Environment.

Australian-based consultancy company, Asia Pacific Waste Consultants (APWC), will be executing the data gathering over the next four weeks, along with the Netherlands-based, Searious Business, and these consultants will be going door-to-door, conducting interviews with households, businesses in the private sector (including hotels) and any other relevant entities who are directly or indirectly associated with plastic waste generation.

Speaking at a press conference yesterday afternoon, Environment Minister Molwyn Joseph explained that the project is part of a larger goal to get Antigua and Barbuda to "zero waste". He implored members of the public to co-operate with the consultants, saying that "what this effort is about, is to give us the data, in order for us to know exactly the volume of waste in Antigua and Barbuda".

The Minister further explained that this data collection exercise will contribute to the completion of the larger project, with the next phase being extensive planning.

"[It will give us the data to plan. The data will help us to plan how we deal with waste, the type of equipment we need [and] to look deeply into how we are going to use this [waste] material," he said.

Project manager with APWC, Martina De Marcos, explained that waste collection and sorting will also be a part of the initial phase of the project.

"Here in Antigua and Barbuda, we're going to be doing interviews, also we're going to be doing waste collection, and then we're going to sort the waste that we collect. So, we're going to be able to know what waste is com-

ing, what type of waste and everything, mainly focusing on plastics, but we're going to be doing all the characterisation of the waste," she said.

The interviews in the data gathering session will begin today from 10am to 2pm, starting in Crosbies, Mount Pleasant, and Royal Gardens, and continue in the afternoon from 3.30pm to 6pm in the Barnes Hill and New Winthorpes communities.

On Thursday, from 10am to 2pm, the interviews will be done in the Piccadilly and Old Road communities, while Grays Farm, Golden Grove and Golden Grove Extension will be done in the afternoon from 3.30pm to 6pm.

References

Antigua and Barbuda Meteorological Service. (2020). *Climate section*. Retrieved 24 November, 2020, from http://www.antiguamet.com/Climate/

Antigua and Barbuda Tourism Authority. (2019). *Antigua and Barbuda receives over one million visitors for 2018*. Retrieved 24 November, 2020, from https://visitantiguabarbuda.com/wp-content/uploads/FOR-IMMEDIATE-RELEASE-Antigua-and-Barbuda-Welcomes-Over-1-Million-Visitors-in-2018.docx

BirdLife International. (2020). *Important bird areas factsheet: Hanson's bay – flashes*. Retrieved 25 November, 2020, from http://www.birdlife.org

Devine, B., Drayton, N., Lindsay, K., Thomas, T., Cooper, B., & Nieves, P. (2010). *Assessment and mapping of Antigua and Barbuda ecosystem resources and promoting a system of protected areas for Antigua and Barbuda*. Retrieved 25 November, 2020, https://parkscaribbean.net/wp-content/uploads/2014/09/Protected-Areas-Analysis-for-Antigua-and-Barbuda-2010.pdf

Francis, S. K. Y., Higano, Y., Mizunoya, T., & Yabar, H. (2015). *Preliminary investigation of appropriate options for Leachate and Septage treatment for the Caribbean Island of Antigua*, in Collected papers for presentation in the 52nd annual meeting of the Japan section of the RSAI, Okayama, Japan. Retrieved 24 November, 2020, http://www.jsrsai.jp/Annual_Meeting/PROG_52/ResumeD/D02-4.pdf

Gore-Francis J., & Ministry of Agriculture, Housing, Lands and the Environment (2013). *Antigua and Barbuda SIDS 2014 progress report* (July 2013). Retrieved 24 November, 2020, https://wedocs.unep.org/bitstream/handle/20.500.11822/8437/Antigua_and_Barbuda.pdf?sequence=3&isAllowed=y

IPBES. (2019). *Global assessment report on biodiversity and ecosystem services of the Intergovernmental Science-Policy Platform on Biodiversity and Ecosystem Services*. Bonn: IPBES Secretariat.

Laws and Regulations

Environmental Protection Act, Antigua and Barbuda, No. 10 of 2019.

The Disaster Management Act, Antigua and Barbuda, No. 13 of 2002.

The National Solid Waste Management Authority Act, Antigua and Barbuda, No. 6 of 2005.

The opinions expressed in this chapter are those of the author(s) and do not necessarily reflect the views of UNU-IAS, its Board of Directors, or the countries they represent.

Chapter 13
Synthesis: Conception, Approaches and Strategies for Transformative Change

Maiko Nishi, Suneetha M. Subramanian, Himangana Gupta, Madoka Yoshino, Yasuo Takahashi, Koji Miwa, and Tomoko Takeda

Abstract This chapter synthesises major findings from the eleven case studies from different countries across the world (i.e. Kenya and Madagascar from Africa; Chinese Taipei, India, Nepal and the Philippines from Asia; Italy, Spain and UK from Europe; Antigua and Barbuda and Colombia from Latin America) concerning SEPLS management in relation to transformative change. It distils key messages in regard to how to understand, assess and take action on transformative change. Implications for science, policy and practice, as well as interfaces between them, are drawn out to address the following questions: (1) what is transformative change? (2) how do we know if we are moving towards a sustainable society? and (3) what are challenges, opportunities and "seeds of change" in the SEPLS context to bring about transformative change? The chapter concludes with five common principles identified across the case studies, while revising the notion of transformative change to reconceptualise it as a radical change that is built on niche innovations of local initiatives and can be fostered through adaptive co-management in the SEPLS context.

Contributing authors to this chapter include Jasmine E. Black, Szu-Hung Chen, Emilio Díaz-Varela, Dixon T. Gevaña, Guido Gualandi, Bishnu Hari Pandit, Andrés Quintero-Ángel, Nathalie Viviane Raharilaza, Krishna Gopal Saxena, Ruth Spencer, Chemuku Wekesa, Debra Williams-Gualandi and Chen-Fa Wu.

M. Nishi (✉) · H. Gupta
United Nations University Institute for the Advanced Study of Sustainability (UNU-IAS), Tokyo, Japan
e-mail: nishi@unu.edu

S. M. Subramanian · M. Yoshino
United Nations University Institute for the Advanced Study of Sustainability (UNU-IAS), Tokyo, Japan

Institute for Global Environmental Strategies (IGES), Hayama, Japan

Y. Takahashi · K. Miwa · T. Takeda
Institute for Global Environmental Strategies (IGES), Hayama, Japan

Keywords Socio-ecological production landscapes and seascapes · Transformative change · Sustainable pathways · Systems approach · Indicators · Landscape governance · Equity · Capacity development

13.1 Key Messages for Transformative Change in the SEPLS Context

Building on the case study findings, we first revisit the concept of transformative change to redefine it in the context of SEPLS management. As a way forward in facilitating transformative change, we then suggest approaches and methodologies to be employed for assessing and gauging progress in moving towards a sustainable society. Finally, we explore strategies to bring about transformative change through SEPLS management by examining challenges and opportunities encountered in the process of facilitating and achieving such change.

13.1.1 What Is Transformative Change?

While there is a general consensus in policy forums on what positive transformative change entails (IPBES 2019; also refer to Chap. 1 of this volume), pragmatic notions of this concept and how it may be attained are expressed at the level of SEPLS. This is because, at the level of operation, stakeholders focus on optimising human well-being and ecological integrity within the landscape or seascape they operate and use resources from and, further, strive to negotiate between local and global priorities of development and sustainability. This means that when 'business as usual' pathways do not seem to provide the desired benefits, actors look for feasible solutions and potential pathways that would enable them to achieve their aspirations. It then becomes an endogenously led, participatory process requiring diverse approaches and capacities towards securing sustainable outcomes that include benefits for the population, economy and for the environment. So, the question that arises is what then constitutes positive transformative change at the level of SEPLS? Based on our experiences, we have characterised transformative change into broad dimensions that may be measured (tangible), and those that are qualitative (intangible), fully acknowledging that there are instances where the boundaries between the two remain fuzzy. Furthermore, it is noteworthy that the tangible and intangible dimensions relate both to social and ecological aspects of the socio-ecological system.

Tangible Dimensions of Transformative Change These refer to aspects that can be physically observed and quantitatively measured. They involve visibly radical changes for the better in practices, approaches, strategies and policy design and implementation relating to the management of SEPLS. Tangible aspects of transformative change could include, for instance:

- Sustainable use and management of natural resources that includes actions to preserve and enhance biodiversity by increasing cultivated species and a higher likelihood of survival and augmentation of native species. This also includes improvements in landscape design and management, soil and water conservation and other environmental qualities resulting in healthy landscapes and seascapes across multiple environmental services. Further, it includes enabling access to a diversity of resources for food, fuel, health and other well-being requirements. This is a common theme across all the case studies, especially in Chaps. 2, 3, 5, 7, 8 and 9;
- Reduced/avoided wastes (e.g. of food, metals, and plastics; resources) (see for instance Chaps. 6 and 12);
- Increased and diversified sources of income (Chaps. 3, 5, 6 and others);
- Calibrated and integrated development plans that are implemented in an interdisciplinary manner to promote sustainable landscape and seascape management, including restoration and regenerative activities (Chaps. 4, 5, 7, 9 and 11); and
- Reduced and avoided negative trade-offs between different socio-ecological components and functions by encouraging diversity (that allows multiple functions to thrive in the SEPLS) and equitable socio-economic transactions between actors (respectful of plural values that exist amongst the stakeholders).

Intangible Dimensions of Transformative Change   These aspects refer to desirable changes in qualitative dimensions relating to the socio-ecological system, and could include, for instance:

- Strong links to individual and collective identities that comprise sense of place, connection to nature, contextually developed agricultural and (innovative) production practices, establishing an emotional and culturally-sensitive affinity to the landscape (see for instance Chaps. 9 and 10).
- Increased awareness, motivations and capacities of local communities, policymakers, and other multiple stakeholders to understand and address unsustainable practices through specific education and knowledge-building activities (Chaps. 3, 5 and 11).
- Improvement in vertical (e.g. between government bodies and communities and their representatives) as well as horizontal (e.g. between community members) communication strategies, efforts and linkages.
- Changes to institutional approaches towards encouraging local/bottom-up contributions and facilitating people-driven processes of landscape management (Chaps. 8 and 9).
- Individual motivations geared towards sustainable practices. This would imply enhanced commitment to priorities related to the diversity and integrity of a landscape and the orientation of cultural values, beliefs and practices towards such transformation (see for instance Chap. 6).
- Anticipatory governance approaches fostered to ensure resilience of socio-ecological systems. This implies cultivating abilities to quickly respond and adapt to ecological changes, that may include climate disturbances and other

environmental changes, and to social, economic or political circumstances (Chaps. 4, 5 and 6).
- Development priorities determined in an inclusive, participatory and endogenously led manner by local communities and other bottom-up stakeholders. This implies active multi-stakeholder participation and collaboration at multiple levels, sense of ownership of management outcomes, appreciation of cultural and religious values, and enhancement of knowledge and well-being of communities. It also implies that governance systems focus on designing and implementing policies that mainstream conservation-related activities (including restoration and regeneration of resources and ecosystems) into development, and actively reduce trade-offs between actors and their preferred activities, and consequently also increase equity amongst them (see for instance Chaps. 3 and 9).

Systems Approach Towards Transformative Change Building on the concept of the socio-ecological system, achieving transformative change is possible when activities take cognisance of the interlinkages between social and natural systems at the level of SEPLS. This could include:

- Extending SEPLS planning and management to the three dimensions of sustainable development[1] including economic (e.g. livelihoods, income, entrepreneurship, and alternative economic models), ecological (e.g. ecological processes, biodiversity, and ecosystem functions) and social (e.g. governance, culture, and property rights).
- Amplifying/increasing synergies between multi-functions of landscapes (socio-cultural, economic and ecological) and minimising/managing trade-offs and tensions across local/regional levels of stakeholders and sectors. This will also mean acknowledging and improving the relationship between ecosystems and people. It also implies enhancing ecological connectivity as an essential component of biodiversity conservation.
- Enhanced networks and partnerships between and among stakeholders at multiple levels across different relevant sectors to support local actions (through mobilising financial, political, technical, and other resources).
- Strong inter-generational links that facilitate the transfer of knowledge, skills and wisdom from the elderly to the youth.

Leverage Points that Enable Transitions Towards Transformative Change
Movements towards desirable states of socio-ecological well-being from an unsustainable one need to be catalysed by several factors. We identify such leverage points of change that can again be characterised as direct and indirect.

1. *Shallow leverage points of change* have a direct or immediate impact on the process of transformative change. Within the socio-ecological system, these

---

[1]We subscribe to the UN definition of sustainable development, that considers development in a more holistic sense than narrow notions of economic progress.

leverage points could involve biophysical or social dimensions and may include the following interventions:

a. Biophysical

- Sustainable land use policies, plans and programmes that include landscape and seascape restoration efforts (e.g. reforestation, agroforestry, and woodlots);
- Management of invasive species that threatens local biodiversity; and
- Conservation of biological resources (e.g. medicinal plants, native species, wild crop relatives, and good soil), and water and biodiversity management practices (e.g. as a resource for agricultural production, traditional agriculture, integrated pest management, and traditional practices of multiple use of biodiversity).

b. Social

- Broadening awareness on the importance of landscape quality, availability of ecosystem services and need for biodiversity protection—this may involve exercises in long-term visioning of the state of the SEPLS;
- Enhancement of food security (e.g. diverse crops, ecological farming methods, etc. that reduce vulnerabilities to natural and economic shocks);
- Enhancement of health security, through better access to medicinal, nutritional and cultural resources, healthier environment and expertise;
- Enhancement of mitigation and adaptation capacities to deal with natural hazards;
- Enhancement of economic benefits and financial incentives (e.g. for entrepreneurship activities), livelihoods, income, and employment;
- Equitable sharing of economic benefits to ensure that all involved in any economic activity are appropriately compensated for their contributions;
- Clarity of roles and responsibilities in resource management among stakeholders, including sole or co-owned responsibilities (such as between governments and communities);
- Financial, technical and other forms of support for upscaling of best practices;
- Avoiding negative incentives and promoting positive incentives (e.g. eliminating subsidies for chemical fertilisers and promoting multifunctional production processes);
- Promoting investments by private sector in the landscape; and
- Setting up accessible knowledge/learning platforms such as online platforms, webinars, peer-learning visits to landscapes that have transitioned to sustainable production, consumption and well-being, farmers field schools, and local awareness programmes in primary and secondary schools.

2. Deep leverage points are dimensions that trigger system-wide change and include:

- Re-orienting perceptions of people towards sustainable and equitable production and consumption. This change requires investing in social learning approaches across stakeholder groups that facilitate or mobilise passion towards the landscape and foster linked values of positivity and curiosity, openness, happiness, and cultural pluralism. Furthermore, fostering values such as reciprocity, equilibrium and collectiveness is important to enhance social cohesion and respectful interactions between different stakeholders. This would enable even the private sector to have a re-articulated vision of their roles and responsibilities within the landscape (as seen in Chaps. 7, 9 and 11).
- Economic drivers: While economic drivers are usually considered to directly influence activities in the landscape as they determine production and resource management activities, they can also be related to intangible aspects (e.g. branding, identity, and fostering dimensions of well-being such as health, access to food, education, sense of place that create a sense of pride and value in the production activities and management of the landscape, as seen in Chaps. 6 and 8).
- Sensitive and adequate facilitation by an external or internal agency for community empowerment, capacity development and enabling communications between different players helps to re-articulate priorities within a community. This can foster meaningful public-private partnerships, and coherent communications to collectively identify needs and strategise on ways forward (as illustrated by Chap. 9).
- Effective enforcement by legal and customary institutions that enable reflexive linkages between policymakers and practitioners (as seen in Chap. 10).
- Education, learning and promotion of knowledge systems that help to re-prioritise attitudes of stakeholders. This should include not just formal systems of learning, but experiential pedagogies that enhance sensitisation/ awareness on best practices of SEPLS management and governance. Importantly, a key point to note is that the traditional knowledge related to SEPLS needs to be promoted (as seen in Chaps. 2, 3 and 6).

13.1.2 How Do We Know If We Are Moving Towards Transformative Change for Sustainability?

Multi-dimensionality makes the monitoring and evaluation of transformative change as challenging as it is complicated. Monitoring the progress in transformative change, therefore, needs to be structured, keeping in mind the diverse set of indicators which may be specific to a landscape and seascape involving multiple ecosystems. It will help to pick the strongest drivers, frame improvement strategies, and explore methods to monitor more qualitative aspects like a community's well-being and ethics.

In this section, we discuss various approaches for monitoring and evaluation (M&E) of transformative change in the context of SEPLS and methods that have been used on the ground to measure it. This helps us capture whether we are really moving towards transformative change. If not, what are the major challenges for M&E? The approaches emerge directly from the experience of the practitioners of the on-ground projects and initiatives.

Volume 5 of the Satoyama Initiative Thematic Review (SITR-5) clearly highlighted the need for M&E of community-based projects. It also highlighted that the valuation methods should be multi-dimensional and multi-faceted, integrating methodologies where necessary, and drawing from multiple data sources over time to provide more comprehensive assessments and contextual explanations (UNU-IAS and IGES 2019). This also includes identifying key performance indicators/goals and ensuring that the interests of various actors are balanced, and local priorities and international goals are coherent. Here, we describe the steps and methods towards real-time monitoring for SEPLS management from the community lens, promoting their inclusiveness and equal participation in the process.

Monitoring and Evaluation Methods
Setting up a baseline is a prerequisite for defining evaluation indicators. Depending on the type of project, there could be one baseline representing on-the-ground conditions before the project or two baselines reflecting the current condition and a control condition. Evaluation indicators should take into consideration the project-specific goals and objectives. This will aid in effective monitoring as the data relevant for the selected indicators will be continuously gathered. The next step, evaluation, helps in assessing the progress in terms of the outputs, outcomes and impacts of the project.

Figure 13.1 shows the detailed steps towards M&E. M&E helps improve accountability and increases the capacity of beneficiaries and implementing staff and partners (SOAS 2013). It also helps in identifying the weak areas and tweaking the system to improve the overall performance and outputs. The evaluation is highly dependent on good quality data collection which must take place from setting the baseline stage until the end of the project, continuously or intermittently. The data collection can be done directly with the help of the community or other stakeholders, depending on the chosen approach as described in the following paragraph.

For M&E, two things must be thought of in the beginning—processes and tools. The on-the-ground experiences presented in the case studies of this book show that various approaches can potentially ensure long-term M&E processes (Fig. 13.2), which include:

Fig. 13.1 Steps towards
M&E

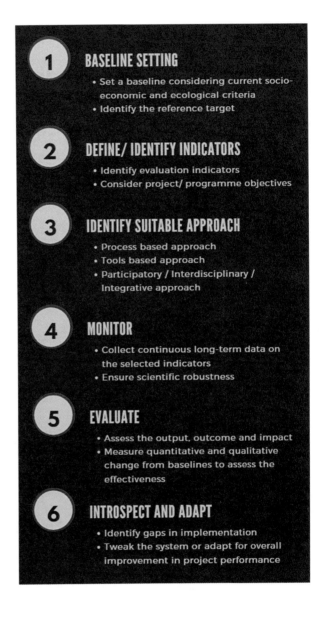

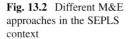

Fig. 13.2 Different M&E approaches in the SEPLS context

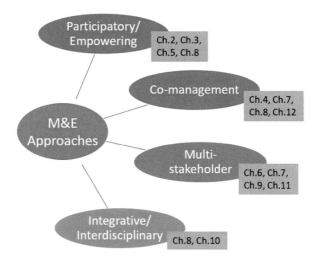

- Participatory and empowering approaches: Local and indigenous communities directly engage in the implementation activities, thereby enhancing their capacities to not only act as the agents of change but also monitor physical progress as they start to understand the baseline and the expected outputs of the intervention. In addition to being able to observe and measure the change based on modern monitoring techniques (e.g. photography, and surveys), they may also use their indigenous knowledge specific to the project area, thereby further enhancing the authenticity of the data collected and the community resilience outcomes of the project. This makes an intervention more sustainable. The data collected directly from the ground is richer and intervention specific. Since the community in this case helps in the successful implementation of the intervention, they can also be motivated to act on the findings. The toolkit "Indicators of Resilience in Socio-ecological Production Landscapes and Seascapes (SEPLS)" provides tools for engaging local communities in adaptive management (Bergamini et al. 2014).
- Co-management approach: This approach overlaps with the participatory approach, but in this case, not only the implementation and monitoring, but also the planning of the intervention is based on inputs from the community. The community is also a decision-maker and chooses the type of intervention based on its needs, and hence helps in co-managing the project. Community engagement from the initial phase also ensures its long-term commitment to the intervention.
- Integrative and interdisciplinary approaches: From the SEPLS perspective, interdisciplinary approaches are also important as they integrate expertise across different levels and sectors to pursue an integrated assessment of land/seascapes. They may combine biophysical and social knowledge to investigate the interlinkages/relationships for effective evaluation of the outcomes. This may result in co-learning where a community's local and indigenous knowledge can

be shared in an inter- or transdisciplinary environment, and other stakeholders can pick cues from that to scale up or replicate the model.

- Multi-stakeholder approach: This approach engages several stakeholders and is sometimes considered as one of the best methods, not only for M&E, but also for achieving the desired goals of the intervention. For this to happen, the needs of different stakeholders and synergy points are identified, on the basis of which their engagement in M&E processes is ascertained. The stakeholders monitor progress from their own perspectives which can be merged to offer a holistic picture. These stakeholders may also be from interdisciplinary backgrounds offering more insights for the successful implementation of the project.

For most of the above approaches, multi-directional communication is a pre-requisite as it is more holistic involving multiple stakeholders. The most suitable approach for a project will depend on the project site, needs and present capacities of different stakeholders. Once the approach for the M&E process is decided, tools come next, which are mainly meant to collect data relevant for M&E. In the context of SEPLS, there are many methods and tools for this process. Tools may vary depending on needs. For example:

- Baseline: Both secondary and primary datasets can be used to determine pre-project conditions, including documents and photos. Secondary data may not reflect on-the-ground real situations and so it is important to check maximum authenticity.
- Measurement of shifts/changes in social norms like community's perception of well-being, qualitative changes in people's livelihoods, fulfilment and enjoyment and connection between people and nature: This can be done by interviews/ surveys, Focus Group Discussions and psychological assessments. This helps in measuring qualitative aspects.

Methods for data collection include:

- Triangulation method: Data may be collected from multiple sources to cross-validate and enrich findings.
- Artistic methods: These include drawing on socially engaged or participatory arts, photovoice (elicitation of visions through photographs taken and evaluated by local people), and storytelling.
- Technological methods: GIS-based spatial mapping or GPS mapping, and smartphone apps (site specific mobile apps for collecting and archiving the data/information) can be used. A drawback for spatial mapping is that it may miss out on qualitative information. To solve this problem, Public Participatory GIS (PPGIS) can be developed using analogic technologies like printed maps, or 3D landscape models made up from cardboard or clay. Also, qualitative approaches like interviews with stakeholders held on-site while visiting specific locations of the SEPLS may produce spatial-explicit qualitative data (e.g. perceptions, values or even emotions elicited while visiting concrete sites in the SEPLS).

Challenges in Measuring Progress

There are many pitfalls in M&E. When we talk about transformative change, which is a relatively newer term, effective processes need indicators from the transformative angle. Of the tangible and intangible aspects of transformative change (see Sect. 13.1.1), tangible parameters are easier to measure as compared to the intangible ones, as they are more quantifiable. In contrast, intangible ones are difficult to be calibrated as putting values on intangible facets may leave out non-numerical values. In addition, each project/intervention has some site-specific characteristics which demands a more tailor-made M&E framework across different spatial and temporal scales. This complex, adaptive character in the case of most SEPLS, is a challenge. Any change may cause adaptation of the whole system or some of its specific parts, and implicitly modify reference parameters for "progress". For example, a project aimed at reclaiming a dump site (progress: clean site) could lead to the recognition of lack of sensibilisation of local inhabitants (progress: awareness), refocusing the effort to promote circular economies (progress: local economies). Some specific challenges include:

- Capturing multi-functional benefits (both tangible and intangible) or multi-dimensional outcomes that are often locally specific.
- Capturing diverse perceptions and preferences on the level(s) of acceptable progress or success among stakeholders. Effects may be realised or felt differently by different communities (e.g. because of caste divide).
- Monitoring multiple parameters like local socio-environmental conditions (e.g. survival of planted trees in a particular region); and benefit sharing protocols.
- Measurement of the effectiveness of resource management.
- Understanding and acceptance of the landscape approaches across different stakeholders.
- Downscaling of global-level indicators (e.g. Aichi Biodiversity Targets, SDGs, and their indicators) is not always appropriate for the project-specific conditions.
- Upscaling of local-level indicators to regional/global levels.
- Insufficient capacity of the communities to use certain technologies needed for monitoring which may hinder their active participation.

13.1.3   What Are Challenges, Opportunities and "Seeds of Change" in the SEPLS Context to Bring About Transformative Change?

Challenges in bringing about transformative change lie in both "enablers" to decide on change and "actions" to implement a decision. The former allows for the latter through rendering conditions, resources, capacities, and means available for actors to make a decision to change their behaviour and then take action for change. Allowing that decisions are made and even formalised in the form of policies, plans and

programmes, transformative change cannot happen unless these decisions are translated into action in practice. While a challenge is making timely informed decisions
at individual and collective levels (including consensus building among multiple
stakeholders), another big challenge is implementation of decisions to take a bold
step for transformative change.

Challenges in Enablers

In the context of SEPLS, agents of change include a variety of stakeholders who
manage SEPLS to derive and use multiple benefits directly or indirectly from across
different sectors and levels. The case studies highlight three major types of
"enablers" that allow for and instigate actions for change: (1) stakeholders' recognition or perspectives on SEPLS problems; (2) their technical and financial capacities; and (3) their authority over SEPLS management. These three types are not
necessarily mutually exclusive, but often interrelated.

First, stakeholders cannot be motivated to change direction or go beyond business
as usual without knowing a problem with current practices of using and managing
SEPLS. Their awareness of problems in question, such as land degradation, water
pollution, and biodiversity loss, is a critical first step in many cases to move towards
a more sustainable society. Yet, as discussed above (see Sect. 13.1.2), it is not easy to
comprehend the problem exactly and evaluate any progress of change due to the
complexity of SEPLS. The SEPLS problems extend to multiple scales (e.g. spatial,
and temporal) and levels (e.g. local, sub-national, national, regional, and international), while entailing multiple and often competing values, perspectives, and
objectives that cut across different sectors. Furthermore, problems stem from
non-linear and dynamic processes of human-nature interactions, involving high-
level uncertainties.

Given the complex and interlinked nature of SEPLS problems, stakeholders often
fail to have a whole picture of SEPLS and recognise the compounding and
multi-dimensional effects of their own practices on SEPLS sustainability from
their standpoints. Urban dwellers are likely to have fewer chances to become
aware of what is going on to maintain the hinterland, while rural-urban migrants
may lose their metaphysical connections to landscapes along with their decreased
physical contact with natural environments. Local practitioners directly engaging in
SEPLS management may not necessarily recognise long-term effects of their practices, and even if their efforts would bring about transformative change in the long
run, they may become easily reluctant to continue without demonstrable, short-term
success within their sight or any interventions to continuously support and encourage such practices.

Secondly, also related to these cognitive aspects, technical and financial capacities of stakeholders can make them better prepared to make a sound decision on
SEPLS management, but inadequacy or imbalance of such capacities often hampers
well-informed and equitable decision-making. SEPLS management technically
requires interdisciplinary knowledge on SEPLS involving social, cultural, ecological
and economic dimensions, but such knowledge is neither necessarily available nor
accessible to all stakeholders. As the case studies show, various kinds of technical

and financial assistance have been offered to increase stakeholders' capacities, ranging from introduction to management models (e.g. supply chain, and multi-stakeholder collaboration), provision of monitoring and evaluation tools, and promotion of awareness raising and public relations to education and training. In recent years, these supports have increasingly attended to continuous and dynamic human-nature interactions within SEPLS. Yet, again as discussed above (see Sect. 13.1.2), interdisciplinary and transdisciplinary knowledge, particularly integrating local and traditional knowledge, is still limitedly available in our society to comprehensively understand SEPLS and adequately evaluate and monitor progress in SEPLS management.

Moreover, these capacity development opportunities are not accessible to all. In particular, the opportunities are not always given to those who are most vulnerable to changes in SEPLS and those who have challenges in having sufficient resources (e.g. time, money, and energy) to address SEPLS problems. Given the context-dependent nature of SEPLS management, participatory and bottom-up approaches have been called for to mobilise and enhance local capacities (e.g. micro-enterprises, and community-based management). In many cases with fragile conditions of national finance, however, political attention is not fully given to finances to support such local and small-scale activities. This is attributed partly to the challenges in recognising the interlinkages between local problems and broader social and ecological impacts and implications of SEPLS management.

Thirdly, authority over SEPLS management is often inequitably granted or unclearly defined to make a decision to move towards a sustainable transition. Stakeholders can justify and legitimise decisions on SEPLS management if authority is given to them to do so, but if not, they can hardly pursue change rightfully. Usually different kinds of authority are granted to different stakeholders as legitimate power to use and manage SEPLS, but power asymmetries exist much in relation to disproportionate technical and financial capacities among stakeholders. For instance, the case of Montespertoli town in Italy points to weaker power of small farmers to control wheat prices compared to large industrialised commercial entities along the supply chain (Chap. 6). With the intervention through the multi-stakeholder efforts, more equitable producer-consumer chains have developed to a certain extent, but otherwise the economic power structure could have been risking wheat biodiversity, amplifying social inequity and further depriving farmers of authority over SEPLS management.

Furthermore, it is sometimes difficult to clearly define rights, roles and responsibilities of stakeholders in managing SEPLS where human and non-human components dynamically interact over time and space. Given the constantly changing contexts and conditions of SEPLS, stakeholders need to repeatedly assert, recreate and exercise their authorities to legitimate their decisions, although such attempts are not always successful (Ahlborg and Nightingale 2018; Sikor and Lund 2009). Dominant power in place (e.g. economic, and political) often supersedes even legislation that encapsulates authorisation over access to and use of natural resources among stakeholders. This leads to *de-facto* immiseration of the legally-protected constituents of society in some cases (Chap. 10) while contributing to continued conventional practices, for instance, through allied opposition against legal change in other cases (Ribot 2009).

To make things worse, all the above three major challenges in enablers (i.e. perceptions, capacities and authorities) can sometimes interplay viciously, serving as a binding barrier to a just and sound decision-making process for transformative change. It is often the case that short and narrow views are dominant among decision-makers who have larger capacities and stronger power to authorise use and management of SEPLS as well as access to resources. Political attention, if it is skewed towards monetary or GDP-based wealth metrics, may foster the vested interests of big business, for instance, for the sake of tax revenues or myopic economic growth. It can thus easily ignore or underestimate local and traditional knowledge and cultural practices that are valuable for SEPLS management but not directly measured with monetary values, consequently hindering transformative change. As such, despite the great potential of political will to make transformative change happen, policymakers and other decision-makers with these perspectives tend to prefer remaining with the status quo and not making a leap forward in fear of slowing or deviating from economic growth.

Challenges in Actions

In the context of SEPLS where decision-making occurs across multiple levels (e.g. local, regional, national, and international) and scales (e.g. spatial, temporal, and jurisdictional), decision-makers may not necessarily be identical with actors who take action on what has been decided. In fact, even when a decision is legitimised and formalised by authorities as legislation or policy, decision-makers most often need to delegate various tasks to others for enforcement and implementation. Such circumstances require alignment of different needs and interests as well as consolidation of different capacities among stakeholders to effectively turn a decision into action. However, divergent levels of commitment to the decision are almost inevitable across different stakeholders who have multiple and often competing needs, interests, views and objectives to manage SEPLS. Likewise, varied levels of engagement and participation in activities based on the decision are often unavoidable given the different technical and financial capacities among stakeholders, though through collaboration they can complement each other to fill a gap in capacities.

In particular, local stakeholders' participation and engagement are indispensable for making change happen, given that they steward SEPLS on the ground in most cases. Nevertheless, they often cannot afford to do so on their own due to their limited technical and financial capacities, even if granted nominal authority over SEPLS management. At the same time, they can hardly dare to be fully committed to a decision that might be associated with high-level uncertainties, being most vulnerable to SEPLS changes in many cases. Most often it is indeed hard to make a bold decision and take a brave leap forward. The case study of the Colombian Pacific region offers a prime example of the predicament of local stakeholders (Chap. 10). Even with the legislation established by the national authority to recognise and protect their collective ownership rights to the lands and to promote their social and economic development, Afro-descendant communities have continuously suffered from illegal exploitation by unauthorised but armed groups, whereas local stakeholders including local authorities had limited technical and financial capacities. In particular, public funds appeared to be insufficient in implementing and monitoring local management plans through local authorities in this case.

Opportunities in SEPLS for Transformative Change

If viewed from the flip side, most of the above challenges can be opportunities for transformative change. In the context of SEPLS, diverse stakeholders have been brought together to decide and act on SEPLS management, taking advantage of knowledge and practices that have been locally accumulated through long-term human-nature interactions. The case studies show the potential of SEPLS to lead to transformative change particularly in regard to the following aspects of SEPLS: "diversity", "wisdoms", and "integrity".

- *Diversity*: Involvement of diverse stakeholders tends to yield mismatches, discrepancies or even conflicts between them in terms of their interests, needs, perspectives and objectives. At the same time, however, such "diversity" renders diverse enablers (e.g. knowledge, financial resources, and alternative means and tools) available to help address shocks, uncertainties and non-linear changes. This as such helps to manage risks and facilitates ecosystem-based adaptation to environmental changes including climate change and pandemics.
- *Wisdoms*: Local stakeholders are often most vulnerable to SEPLS changes but have accumulated and enriched practical and experiential knowledge on SEPLS management, nurturing local and indigenous knowledge. Their knowledge has been embedded in local contexts as "wisdoms" for their survival but could be applied and extended through bottom-up approaches (e.g. local empowerment, and peer-learning) to promote a more sustainable and resilient society.
- *Integrity*: Dynamic interactions between human and non-human components across different levels and scales disallow a one-sided solution to address multiple dimensions of a SEPLS problem. Yet, vertically and horizontally "integrated" strategies help increase synergies and minimise trade-offs between different elements of SEPLS to effectively meet multiple needs for improved human-nature relationships. This can be made through mobilising diverse resources and capacities of stakeholders and facilitating multi-stakeholder collaboration with involvement of youths, women and the elderly.

Moving from Seeds of Change to Transformative Change

Most of the cases exemplify "seeds of change" within which transformative change has been emerging as observed in some positive outcomes for local communities or diffusive effects to surrounding or different regions (e.g. replication of good practices), but is yet to fully happen. Importantly, some niche innovations (e.g. sustainable farming practices, equitable supply chains, and buy-in of business and industries in recycling) have been already put in place largely at the local and regional levels in most of the cases in this volume. These innovations, however, are still in 'incubation rooms' somewhat waiting for radical changes (Geels 2005, p. 684). To take a further step in bringing about transformative change, a more systemic approach would be required to address deeper leverage points (i.e. the places for interventions in a system, which are deeply rooted in the causes of unsustainability) rather than shallow ones (i.e. the places for easily implementable interventions) (Abson et al. 2017).

In connection to deeper interventions, the findings from the case studies suggest that a more strategic approach to SEPLS management would help facilitate turning "seeds of change" into transformative change, particularly in the following respects: (1) concept of values or mindset; (2) governing processes; and (3) governing outcomes. The first point is relevant to the "intent", one of the system characteristics of SEPLS in the realm of deep intervention, underpinning values, goals and worldviews of actors that shape the direction of change. The second and third ones are relevant to the "designing" characteristic of SEPLS also in the realm of deep interventions, which determines social structures and institutions to manage shallow interventions (also see Chap. 1).

Concept of Values or Mindset First, to break the status quo, stakeholders' conception of values or their mindset should be diverted from what is skewed to economic growth based on the currently entrenched metrics such as GDP to what is extended to multiple facets of well-being. As the case studies elucidate, values of SEPLS are reciprocal across different domains of SEPLS (e.g. circular economies) rather than linearly cumulative, entailing multiple benefits for well-being (e.g. health, and quality of life) (e.g. Chaps. 2, 3, 5, 6, 7 and 12). This can be demonstrated in several ways as follows:

- Communicating multi-dimensional values of a certain product to its consumers would help change their consumption behaviour (e.g. Chaps. 6 and 7). It may also effectively promote chains of changes by taking advantage of existing supply chains but ensuring equitable sharing and distribution of benefits.
- Innovative educational and capacity development approaches would help people to recognise themselves as agents of change through identifying their rights, power and contexts particularly in connection with their individual or collective emotional links to the natural environment (e.g. ethno-education, visioning exercise, and peer experiential learning, for instance, demonstrated in Chaps. 8 and 10), instead of top-down or one-way knowledge transfer. They can facilitate behavioural changes and foster better relationships between humans and natural environments.
- Efforts to raise awareness of biodiversity at every chance, for instance, through daily life, social media and campaigns, would help promote biodiversity mainstreaming among diverse stakeholders. The target audience would range from lay people to politicians for mainstreaming in policymaking and implementation as well as from youths to the elderly for inter-generational equity and empowerment of future generations.
- Multi-stakeholder and participatory learning processes would help identify innovative means for more sustainable human-nature relationships (often bringing in

various types of knowledge including local, indigenous and scientific ones) and scale up good practices (e.g. Chaps. 7 and 9). For example, groups of farmers experiment with sustainable methods for which field days and discussions are organised with involvement of various stakeholders to scale it up for the wider community.

Governing Processes Secondly, governing processes should be inclusive and participatory where all stakeholders can have a say in what SEPLS could be sought after and how SEPLS should be achieved and managed, allowing for inclusion of multiple values held by diverse stakeholders who act on decisions as agents of change. At the same time, the processes should ensure that stakeholders finally legitimise the decisions made in the processes for authorised implementation and enforcement to make transformative change happen. To do so, it is crucial to exploit all opportunities from across different levels, scales and sectors of SEPLS management all along with the participatory governing processes including policymaking and implementation. This would enhance communications, facilitate mutual learning and help address inequalities between all stakeholders. It would further allow for better chances for effective and equitable multi-stakeholder collaboration and promote ownership of the SEPLS by the local communities. Some examples include the following, but all together would facilitate the processes to incite a mass movement towards a sustainable society:

- A hybrid of top-down and bottom-up approaches to policymaking would contextualise local problems in policies and vice versa and render longer-lasting effects through fostering ownership of decision outcomes among multiple stakeholders (e.g. Chaps. 7 and 8).
- With a view to the governing process including policymaking and implementation as a learning opportunity, involving government authorities from the early stages, possibly from multiple levels, would facilitate political buy-in and ensure authorisation of decision-making outcomes (e.g. Chaps. 7 and 12).
- Through policy cycles as learning processes, good practices can be scaled up and out and deepened to ensure knowledge-policy-practice linkages. Inputs into policy cycles can be made, for instance, through interactive workshops among stakeholders (particularly involving policymakers) and brief and easy-to-digest publications for policymakers.
- Along with the processes, facilitators serve as key agents of change to identify a common language for conversation, create communication channels, mobilise technical and financial capacities, and secure representation among all the stakeholders (e.g. Chaps. 2, 3, 4, 5, 6 and 9).

Governing Outcomes Finally, governing outcomes resulting from the governing processes should build in reciprocal, equitable and interactive connections between human and non-human entities. Governing outcomes should not be static but rather dynamic to flexibly adapt to change when and where appropriate, but a certain governing structure could be aimed for to make governing efforts sustainable. Within the governing structure, rights, roles and responsibilities of stakeholders should be clearly defined with transparency. Some ways forward include:

- To take advantage of scientific evidence as well as practical lessons learnt from experiences, science-policy-practice linkages should be integrated in the governing structure. For this purpose, networks of diverse actors including indigenous peoples and local communities may pragmatically serve to transmit traditional knowledge and give voices to those who manage SEPLS on the ground.
- Equitable production and supply chains should be entailed in the structure finally to facilitate behavioural change of consumer (i.e. end users) for sustainability.
- The working of the governance system should be accountable and transparent to all the stakeholders. This allows them to continuously legitimate and justify SEPLS management, knowing who exercises what authorities to provide what benefits to whom. It would also help to identify opportunities to improve and enhance multi-stakeholder collaboration to pursue transformative change.

13.2   Conclusion

Bringing about transformative change towards sustainability in the SEPLS context requires pragmatic approaches to ensuring the well-being needs of the community along with the integrity and diversity of the ecosystems and resources therein. The bottom line is that the priorities of conservation, restoration, sustainable use and equitable sharing of benefits arising from SEPLS need to be integrated into actions by different actors operating in the SEPLS who are motivated to deliberate and choose sustainable pathways. The motivations of those actors are essential for transformative change as they drive such change endogenously and sustainably. These motivations could be triggered by interventions such as innovative educational and capacity development pedagogies that instil a sense of place and pride in engaging in sustainable action; and participatory and interdisciplinary approaches to identify challenges and solutions in managing the SEPLS to enhance a sense of ownership and follow-up actions amongst all actors. Such (deep) interventions can be designed to bring about a systemic, society-wide change but would take a longer time than immediate fixes to show results.

Other interventions that have impacts include, for instance, policies that support sustainable production and consumption, reduce waste, promote recycling, invest in conservation of crop diversity or ecosystem services amongst others. These interventions are not trivial as they help make endogenous SEPLS management more persistent and legitimate. As the case studies show, the SEPLS management exemplifies the "seeds of change" manifesting niche innovations with great potential to lead to a regime shift in bringing about transformative change. Depending on the contexts, requirements and information at hand, various intervention models have been explored across different sites. These models are highly context-dependent and can be strengthened through the collaborative processes where the stakeholders engage and negotiate in identifying optimal solutions amongst different and sometimes competing needs and interests.

While distinct in their approaches depending on specific contexts and circumstances, the case studies of SEPLS management also underscore several common principles, including the following:

- *Endogenously driven actions based on value pluralism*: Actions need to be endogenously driven by different stakeholders at the community level who proactively design their plan of action and make and implement decisions to address their needs and interests. In determining their decisions and actions, all stakeholders' perspectives should be respected and the plurality of values should be acknowledged in relation to the use and management of resources as well as the development pathways that exist among the actors.
- *Systemic and transdisciplinary approaches to fostering niche innovations*: A systemic approach should be taken to foster niche innovations in managing SEPLS and governing all resources within the landscape and seascape. This requires creative integration of different knowledge systems to arrive at sustainable solutions and therefore the expertise of people with a diversity of backgrounds (farming, education, capacity development, conservation, value addition, etc.). In particular, the approach needs to be reflected in the implementation of national policies to ensure coherence between multiple policy objectives as they interact so that they are mutually supportive across different sectors.
- *Equitable authority over SEPLS resources*: In governing the SEPLS management, authorities over access to and use of natural resources should be equitably granted and clearly defined for policymaking and implementation to move towards sustainable transition. In this regard, being respectful to customary rights and local priorities is a necessary condition to ensure that both social and ecological goals are achieved.
- *Coordinated multi-level networking through peer learning*: To allow for a systemic approach, networking between actors at multiple levels and scales is important, and often benefits from having a strong facilitator to convene and mobilise the community towards envisioning and acting on a common agenda. Related to networking is the importance of fostering peer learning within and between communities and other actors in and beyond SEPLS. It enables more effective sharing of learning experiences regarding contextually replicable solutions towards sustainability.
- *Iterative participatory and inclusive assessments for strategically steering transitions*: Developing a clear strategy is needed to monitor, evaluate and adaptively manage changes towards desired outcomes so as to take steady steps for sustainable transitions. The process for this should be participatory, inclusive and respectful with clear boundaries that enable systematic assessments and course corrections.

It is noteworthy that these principles, distilled from actions on the ground, speak directly to several conceptual dimensions relating to transformative change that are being discussed in mainstream literature. The case studies along with this synthesis make the point that experiences from the ground can inform policymaking processes to enable better policy design and more effective implementation to meet policy objectives. Furthermore, we make a strong argument that transformative change is a concept that is already being explored on the ground, and even if it may be variously interpreted depending on socio-ecological contexts, there are some clear principles that emerge and can inform its uptake at multiple levels of implementation.

Revisiting the concept then, transformative change in the SEPLS context may be conceptualised as a radical change that is built on niche innovations of local initiatives and fostered through adaptive co-management and use of a mosaic of ecosystems towards enabling socio-ecological resilience. The process of change is endogenously led and inclusive of the plural values held by different stakeholders in the system, whereas deliberations and negotiations amongst different stakeholders are pursued. Through a systemic approach that also facilitates integration across different levels and sectors, it promotes multiple agendas to fundamentally address local needs and interests but extends to achieving global goals—conservation, restoration and sustainable use of biodiversity and ecosystems, innovative production and governance practices, equitable transactions of natural resources among actors, creative integration of multiple worldviews and knowledge systems through peer learning, continuous monitoring and evaluation of progress, and multi-stakeholder partnerships for collaborative actions.

References

Abson, D. J., Fischer, J., Leventon, J., Newig, J., Schomerus, T., Vilsmaier, U., von Wehrden, H., Abernethy, P., Ives, C. D., Jager, N. W., & Lang, D. J. (2017). Leverage points for sustainability transformation. *Ambio, 46*, 30–39.

Ahlborg, H., & Nightingale, A. J. (2018). Theorizing power in political ecology: The "where" of power in resource governance projects. *Journal of Political Ecology, 25*, 381–401.

Bergamini, N., Dunbar, W., Eyzaguirre, P., Ichikawa, K., Matsumoto, I., Mijatovic, D., Morimoto, Y., Remple, N., Salvemini, D., Suzuki, W., & Vernooy, R. (2014). *Toolkit for the indicators of resilience in socio-ecological production landscapes and seascapes (SEPLS)*. Rome: UNU-IAS; Biodiversity International.

Geels, F. W. (2005). Processes and patterns in transitions and system innovations: Refining the co-evolutionary multi-level perspective. *Technological Forecasting and Social Change, 72*, 681–696.

IPBES. (2019). *Global assessment report on biodiversity and ecosystem services of the Intergovernmental Science-Policy Platform on Biodiversity and Ecosystem Services*. Bonn: IPBES Secretariat.

Ribot, J. C. (2009). Authority over forests: Empowerment and subordination in Senegal's democratic decentralization. *Development and Change, 40*, 105–129.

Sikor, T., & Lund, C. (2009). Access and property: A question of power and authority. *Development and Change, 40*, 1–22.

SOAS. (2013). Project planning and management-C134 unit ten: Monitoring and evaluation.

UNU-IAS & IGES. (2019). *Understanding the multiple values associated with sustainable use in socio-ecological production landscapes and seascapes (Satoyama initiative thematic review)* (Vol. 5). Tokyo: United Nations University Institute for the Advanced Study of Sustainability.

Printed in the United States
by Baker & Taylor Publisher Services